CONTENTS

Fire, Flint and Faith

Xavier Suarez

authorHOUSE®

AuthorHouse™
1663 Liberty Drive
Bloomington, IN 47403
www.authorhouse.com
Phone: 833-262-8899

Published by AuthorHouse 08/05/2022

ISBN: 978-1-6655-6713-8 (sc)
ISBN: 978-1-6655-6714-5 (e)

Print information available on the last page.

This book is printed on acid-free paper.

PROLOGUE

The celebrated Netflix series, *THE CROWN*, features a marvelous scene in which Britain's Prince Philip faces his long-estranged mother, Princess Alice of Battenberg. As a child, this royal lady had been wrongly diagnosed as deaf and schizophrenic.

Later in life, she managed to survive some form of barbaric therapy to become a nun and to found a new order dedicated to saving Jews during the Holocaust and to helping the poorest of the poor.

Mother and son are reunited in 1967 at Buckingham palace, a refuge given to her by her royal family in order to remove her from the violence taking hold of Greece at the time. In the featured scene, she questions Prince Philip as to his faith; he acknowledges that it is shaky, at best.

Then this former princess-turned-nun says to him: "Find yourself a faith. It helps. Not just helps. It's everything."

The story of humanity is often told as a tug-of-war between two extremes: one that values faith above all things and one that denies faith in all its manifestations. The latter view is the one that prevails in many of the so-called "scientific" halls of the modern European and American campus.

The former view, which considers faith in God the essence of all things, the key to our happiness and to the full flowering of our humanity – the sum total of the human condition – is not considered "scientific" at all. It is tolerated in the theology or religion department of campus, and occasionally makes its presence known in other departments (psychology, philosophy and sociology), but only to the extent that it helps to explain mysterious cures or unknown correlations between human behavior and faith in an almighty being.

This book is not intended as an *apologia* for religion or faith as an important value, although it will most likely have that effect on my reader. It is more like a survey of what we know about our species, using all of the tools that science provides.

But what exactly is "science"? As I was writing these lines, I was confronted by a fifteen-year-old girl, named Lauren, who seemed enthralled by the question. She said: "I am doubtful of any analysis which leaves out the *wonder* in the human condition. To me, that element is as important as any other, and needs to be part of the search for truth."

Lauren is right. To insist that the scientific search for truth must be limited to things that can be measured with a clock or a yardstick is to exclude relevant evidence that is measured indirectly when we analyze human wellness. Modern psychology is all about measuring those indirect effects.

Mental health is in great part about motivation; and motivation is often an outside stimulus in the equation of human happiness. Or I should say outside *stimuli,* since there are many manifestations of this phenomenon. Humans, as we shall see, are motivated by witnessing and reading about what other humans think and do; they (we) are motivated by song, and poetry and visual arts.

There is also an inner stimulus that seems to come from some part of us that is not defined by our senses. Most of us – perhaps all of us – are motivated by the inner conviction that there may be life after death and that the measure of happiness in that the after-life is integrally related to the way we conduct ourselves during our physical lives.

All of these motivational realities are part of the scientific method. Understood properly, science is what Nobel Laureate Percy Bridgeman called "doing one's damndest to search for truth – no holds barred."

That includes insights that emanate from looking at the human condition as a triune reality, commonly referred to as body, mind and spirit.

Let me add that unlike many philosophers and theologians, I do not attempt to suggest which of the three is more important. In other words, my analysis does not include a bias in favor of spirituality over matter or mind over body.

My analysis presupposes no hierarchy among the three dimensions of the human reality. It simply takes all three into account. The methodology has been referred to as "integral humanism." (My niece, Margarita Mooney, currently a professor at Princeton Theology Seminary, has recently referred to the same philosophical concept as "humanistic synthesis.")

Much of academia oppose this integral way of thinking. They cling to a strictly materialistic explanation, often referred to as "scientific materialism." (I define it, pithily, as "only matter matters.")

In the field of anthropology, the logic of scientific materialism has led some scientists to embrace the idea of what is called "sociobiology." It appeared first in the writings of Harvard's E. O. Wilson and it was not particularly welcomed, for the simple reason that it seems to absolve humans of any culpability for antisocial behavior.

Sociobiology seeks to explain human behavior as a strictly biological phenomenon. It has not prospered much since being introduced, in the 1970's. Instead, we see science taking a more holistic approach.

We see this particulary among health-care professionals, who find the purely materialistic approach singularly ineffective in understanding and treating mental health issues.

That should not surprise us. Describing any member of our species based on biochemistry alone is bound to fail. Diagnosing any illness, particularly those of the mind-psyche, exclusively in terms of the instinct to survive and procreate, is nothing less than professional malpractice.

Humankind is nourished by food, by the companionship of others, and by the image we humans see of ourselves as worthy of the love of our peers and of our own inner selves. Those who suffer much, and those who dream much tend to seek in a higher power the kind of inspiration that is often needed to endure and excel.

And all of us, whether we have suffered much or dreamt big dreams, can only be fully human if we see in all others of our species a basic equality. For most of us, that insight comes from realizing that all humans are sons and daughters of a common creator.

In this book, I argue that the same realization can be deduced from a proper reading of history and from the step-by-step acquisition of knowledge that our species has pursued as we became more and more civilized.

I will argue that advanced civilization requires collaboration, that collaboration requires trust, and that trust requires the kind of faith in human goodness that only comes from belief in a common parent.

When humans share key discoveries, formulas, ideas and inventions, they thrive collectively. Just as two heads are better than one to solve a problem, two societies are better than one at advancing civilization.

Fire and flint were the first two key discoveries. The idea that there is one, common, loving creator was the third one; faith in such a "god" brought people together and enabled them to collaborate in finding truth and shaping a more prosperous society. Faith prompted our ancestors to suspend their prejudices, allay their fears and commune with the tribe next door.

This book thus argues that fire, flint and faith were the key factors that helped end tribalism and begin civilization.

Getting along with other tribes requires, as a first step, that we get along with our own tribe. My good friend, who is presently the archbishop of Boston (Cardinal Sean O'Malley) likes to say that "we must love our enemies as much as we love our neighbors – and, often, they are the same people."

Having neighbors who are enemies is the norm in uncivilized societies. By contrast, making friends of your neighboring tribes is the norm of societies that accept the idea that we are all made in God's image.

That is the story of faith, which is the subject of Chapters VII and VIII. There, as well at various other points in my narrative, I describe how communication and collaboration among tribes began with tolerance for the differences that exist even between members of the same family.

People with the same genes and with the same upbringing are markedly different in their conduct. That reality, which surprises the scientific materialists, is the topic of the next section.

INTRODUCTION

$\mathcal{I}$t has been a week since my cousins, on my mother's side, got together in Miami for a family reunion that takes place every five years. The last name that unites us is "Gaston"; it comes from my maternal grandfather, whose family came from Spain and had some royal ancestors.

On my father's side, the Suarezes were also Spaniards. They were more like the typical immigrants to the New World: initially not particularly well off, but after a couple of generations fairly well-to-do.

At one point, both families owned small sugar mills; however, because they were prolific donors to the poor and the church, and also prolific in child-bearing, there wasn't much of an inheritance by the time I came along, with my thirteen brothers and sisters.

Both of my parents were very religious; and they both instilled in us a reverence for some sort of divine being and also, because we were all "children" of that divinity, for our fellow human beings. As in the two greatest commandments, we always knew that our mission in life – besides fulfilling our own basic needs – was to love God above all things and our fellow human beings as much as we love ourselves.

We had no idea why that was the case; but we accepted it.

Of course, that was what we were taught; it wasn't necessarily what we did in real life. But we did strive to see all members of our species as our equals; we did our best to not feel either insensitive to those who had less than us, nor complacent with their suffering.

I recite this bit of family history to introduce the theme of this book, which is an effort to use modern science to shed light on the mystery of human personality. Let me correct that: not human "personality" but human personalities.

Each Person Is Unique

In Spanish we have a saying: "Cada persona es un mundo." ("Each person is a world unto himself/herself.") No two humans are identical – not even identical twins, as we shall see.

When the Gastons got together with the Suarezes, the differences were pronounced, even in the first generation, where genetics had not exerted its diluting effect from marriages to others.

The Suarezes, in general, were more intense, taller, and darker in skin and eye-color.

But all generalizations fail the moment one focuses on the individual. And most of it is not biological. Most of the personality differences, I will argue, result from choices made.

Of course, to make choices, a human being needs to have control over instincts; people who have serious emotional or mental dysfunctions are often guided by their psyches into doing things that even they don't necessarily approve.

That is the case with addicts.

And we are all addicted to some extent. Even the most mature in our species have biases and some built-in narcissism. Jews and Christians think that is a result of some "sinfulness" on the part of our ancestors. They think it happened when our forefather (Adam) was convinced by our foremother (Eve) to deviate from the proper diet and pollute his body with a forbidden fruit.

The theory is not that different from modern ecological science, which tells us that polluting the atmosphere or the ground in one continent affects the healthy environment of all continents.

The book of Genesis, it appears, may contain the first ecological lesson of history: Eat right or you will suffer the consequences.

But I am getting ahead of myself. The basic point is that nature (biology) and nurture (psychic development of the maturing child) do not complete the picture of the human personality.

Decisions freely made by each of us complete the picture of us. And that applies even to my own fourteen siblings, all of whom are both Suarez and Gaston.

The Fourteen Suarez-Gaston Siblings

Yes, you heard me right; I come from a family of fourteen brothers and sisters, all born of the same parents. We have the same genes. As far as nurture goes, we all went to religious schools in Cuba, professed the same basic Christian religion (Catholic or Protestant), and achieved similar levels of academic success. Eleven of us completed four-year college degrees; one completed a two-year associate degree. About half of the siblings obtained advanced degrees.

Our academic aptitudes were not much different. We all were good in math; we all played chess well – some exceptionally well. We all had above average I.Q.'s, although one had to struggle with various learning disabilities, including dyslexia, as well as a mild case of Aspergers.

So why are we so different? And how much do our differences have to do with evolution, or the instinct to survive? Why do human beings – even those who have absolutely no survival issues, because they were born to wealth – argue so much?

And why do some humans give up the right to procreate and the right to own material goods, in order to dedicate their lives to teaching the Gospel? Why do my younger brothers retire, while I aspire to more professional and occupational challenges?

One time, as I was eagerly awaiting my turn to borrow a best-seller from our public library, I realized that I was the only one of the 14 siblings who would care to read the biography that Robert Woodward had just written about Donald Trump. Another time, I picked up the most recent biography of Winston Churchill (the one by Robert Andrews, who was the first Churchill biographer to obtain access to the king's and queen's diaries) and again wondered why only I was likely to plow through this magnificent account of this amazing statesman.

We all admire Churchill in my family. But why am I the only one who devours every new biography, as if it were a newly written chapter of the Bible?

This book tries to answer those questions. The answers I am formulating will flow from every science I have studied (including anthropology, which I am doing my best to self-teach and psychology, which I have studied in spurts and starts, particularly on issues like bipolar depression, analysis of multiple intelligences, treatment of addiction and holistic medicine.)

Those studies supplement what I have learned formally in areas like engineering, economics, physics, math, political science and law and informally in what are perhaps the two most important of all fields of human knowledge that illuminate questions of ethics: philosophy and theology.

None of those studies, except anthropology, suggest that there is much importance, as far as human behavior, in analyzing how we may have evolved from animals.

Anthropology is different. Here, the conventional wisdom – derived from evolutionary biology – questions whether there is any role for free will in human conduct. And that is logical, since if human traits evolve naturally and spontaneously from the effort to survive and procreate, there is little role for choice.

Curiously, the other sciences, including psychology, spend little time on the question of free will. Professionals who make a living from provoking a change in human conduct (such as the treatment of criminality, depression, uncontrolled anger, narcissism and addiction), simply assume that their science, coupled with great effort by the patient, can bring that about.

But that doesn't keep great minds, particularly anthropologists and evolutionary biologists, from trying to analyze human behavior using strictly evolutionary mindsets and methodology. They are convinced that such narrow methodology will ultimately explain all dimensios of humankind.

As we shall see, that paradigm is deficient – often laughably so.

Now it is time to start our analysis with a bit of natural history. So let's start at the beginning.

My first chapter borrows a page from the famous Stephen Hawking; I also borrow the title to his book, *A Brief History of Time*. My purpose is to condense the story of recent scientific discoveries which point to the existence of an infinite (or near-infinite) force that acted on the universe at various times from its beginning to the point in time when our species arrived on the planet.

The reason for this bit of cosmology and biochemistry is not to bore my reader with fancy scientific concepts. It is rather to prepare you for some startling new insights into the transition from animals to humans.

That transition is the main subject of this book. It helps, in understanding it, to keep an open mind, enough to resist the impulse to assume that humans are just the highest species in the food chain.

We are much more than that. It says here that we are much closer to God.

CHAPTER I
A BRIEF HISTORY OF TIME

Following time backwards, we find more and more organization of the world. If we are not stopped earlier, we must come to the time when the matter and energy of the world had the maximum possible organization. To go further is impossible. We have come to an abrupt end of space-time – only we generally call it the 'beginning.' Sir Arthur Eddington

A surprise from the Deep Field is the universe's lights, contrary to astronomers' hunch, turned on in one great burst. It was as if every chandelier in a mansion were flickered on simultaneously on a moonless night. Sharon Bagley, *Newsweek*, November 3, 1997.

$\mathcal{M}$y brother Mel, who has a Masters from M.I.T. in engineering and also knows a lot of theology, is imbued with the idea that every trait in our species has an evolutionary cause.

Recently, he posed the following question: "What is wrong with believing that God created the universe at the Big Bang and then allowed spontaneous forces to take over from then on - without further interference?"

Well, I answered, there is nothing theologically wrong; many if not most theologians accept the possibility that after the Big Bang, there was

1

only one other "special" creation – when the almighty inserted a soul into what may well have been an animal ancestor of our species. But in doing so, you are ignoring about 99% of the data, which indicates that humans represent a major, unbridgeable break from animals and also indicates that even organic life seems incapable of coming together spontaneously from the energized movement of inorganic particles.

My brother reminds me of the fiction writer who casually offered his impatient view of faith in a divinity – unless that divinity worked through evolving viruses. The writer was named Charles McCarry and his novel was called *Mulberry Bush;* here's the quote that intertwined Darwinian natural selection with the divine mind:

> *I could not believe unless the mind of the creator turned out to be invisible, invincible, omnipresent, immortal bacteria and viruses that collectively drove the evolution of our species over billions of years with the objective of producing an organism intelligent enough to transport the organisms to other planets so that they could begin the process all over again.*

McCarry is well published, well read, and well educated. He writes good fiction. But he has no grasp of real science.

No matter. He clings to any theory that will allow him to continue believing that the Big Banger had to do her creative work in a semi-infinite series of spontaneous, undirected, random micro-events that somehow caused atoms, initially spread all over empty space, to merge into wonderfully complex molecules, then coalesce into tiny living factories that could not only reproduce but ultimately become more complex and end up looking like Lady Gaga and LeBron James.

Darwinism can be described as having two basic truths (ancestral origin of animal species and chronological progression towards complexity) sprinkled with non-sequiturs as to causality. When confronted with what evolutionary biologist Ernst Mayr calls "bridgeless gaps," the hard-line Darwinians simply resort to fantasy.

One of the most impressively illogical of those is what they invent to get around the matter of biogenesis, i.e, the beginning of life. At this point and based on the latest discoveries of science, the biogenesis phenomenon is so incomprehensible, so intractable, so different from what came before and so unlikely to have resulted without outside interference (hence, so violative of the Second Law of Thermodynamics) that it begs for a conclusion that a perfect, or near-perfect being, did the trick.

No self-respecting sociobiologist wants to admit of such a conclusion, no matter how illogical the alternative. Admitting that they don't have a clue is inconceivable for some scientists. And so they invent strange, unprovable, scientifically unsupported theories like "panspermia."

Panspermia

When confronted with the mystery of biogenesis, a scientist or two has been known to become mystical and invent whatever theory pops up, no matter how improbable. Confronted with a likelihood of having to accept a new infusion of order from outside the system – a second Big Bang – the fiercely agnostic scientist turns to science-fiction.

In the case of biogenesis, it is called "panspermia."

In summary, panspermia means that an extra-terrestrial being came to our planet and injected inorganic molecules with a dose of life. Finding no natural causes in existence when cells first made their appearance, the convinced and committed materialist simply invents an "ET" that leaves his or her home in another planet and makes a new home on Earth.

That's not science; that's science-fiction.

We, on the other hand, stick to what our instruments show us about the startling complexity of life. The science of genetics is our guide.

Without getting too technical, we can refer our reader to what Francis Crick (co-discoverer of the key genetic element known as DNA) calls his "Central Dogma." It goes roughly as follows: A DNA produces an RNA and an RNA produces a protein. But a protein cannot produce either an RNA or a DNA.

Crick's "central dogma" is akin to what the layman calls the paradox of what came first – the chicken or the egg.

What Came First – the Chicken or the Egg?

Bill Bryson is the widely published author of *A Short History of Nearly Everything*. He offers us the following description of the riddle posed by the mutual dependence of DNA and proteins:

> *A protein to be of any use must not only assemble amino acids in the right sequence, but then must engage in a sort of chemical origami and fold itself into a very specific shape. Even having achieved this structural complexity, a protein is not good to you if it can't reproduce itself, and proteins can't. For that you need DNA. DNA is a whiz at replicating – it can make a copy of itself in seconds – but it can do virtually nothing else. So we have a paradoxical situation. Proteins can't exist without DNA and DNA has no purpose without proteins. Are we to assume that they arose simultaneously for the purpose of supporting each other? If so, wow.*

Wow, indeed!

And it gets worse – or perhaps I should say better. I suppose it's worse for the agnostics and better for the believers. It turns out that DNA replication and protein construction is not the only complex mechanism going on inside the cell. The entire thing is a marvelously crafted, multi-task, synchronous, well-oiled, perfectly timed and perfectly tuned engine whose components are uniquely designed to carry out functions that never occurred before in history.

Modern scientists marvel at the precision of it all – the extraordinary, unprecedented craftsmanship. They also marvel that naturally occurring elements that preceded the formation of the first cell did not seem conducive to the formation of living cells, even if some invisible hand was able to assemble them in one fell swoop.

The Cell as a Little Factory

One problem encountered by scientists in explaining what appears to be a rather sudden appearance of a living cell, about four billion years ago, is that the atmosphere on earth, as far as we can discern, did not seem suitable for living organisms. I won't test my reader's patience with the biochemical analysis, because there is a much bigger problem.

The bigger problem is that organic life requires the simultaneous appearance of the right components of many other chemicals and organisms in just the right concentrations.

In other words, the Planet Earth is a like a three-dimensional jigsaw puzzle that assembled itself in a very short period of time, using processes that were not known in nature. What I mean by that is that the creation of a self-reproducing cell requires something akin to a little factory; but there was no medium-sized factory to create that little factory. *There were no factories at all on earth prior to the time the first little factory appeared on Earth.*

Here's how Yale's Lyn Margulis describes the process:

> *Some 2,000 to 5,000 genes make a similar number of proteins. Proteins and DNA mutually produce each other within the cell membrane that together they fabricate. Bearing a common biochemistry, all life probably dates to a single perhaps (but not necessarily) improbably historical moment.*

Margulis is somewhat of an exception among modern biologists. The majority hate to admit that they don't have a clue as to certain phenomena.

And so they invent strange, unprovable, scientifically unsupported theories, such as the previously mentioned panspermia. This intellectual juggling act reminds me a lot of the one that fervently agnostic scientists invent to explain the Big Bang itself.

Of Parallel Universes (and Other Fairy Tales)

In order to avoid the generally accepted fact of a Big Bang, there is a school of scientists who have come up with the idea of parallel universes. In short form, the theory is that our universe is not the only one out there; to the contrary, it is one that resulted from the infinite possible combinations of elements that must have always existed, and it is far from unique.

And I mean very, very far from unique. As the theory goes, there are an infinite number of universes and – by the law of probabilities – an infinite number that have someone with my exact physical characteristics. An infinite number of Xavier Suarezes…How cool!

How silly!

It is simply amazing the lengths to which some scientists will go to avoid the simple fact that right before the Big Bang there was nothing: no space or time or matter. After the Big Bang, there was enormous mass, infinite density, and space-time.

And we will never know how it happened. Well, let me rephrase that: We are increasingly able to *calculate how it happened*, by extrapolating with known equations back towards the beginning of time. But we can never know *why it happened, in terms of cause-and-effect.*

We cannot replicate, in a lab, either the Big Bang itself or the fraction of time immediately afterwards, when the universe took shape. But we have some idea of the magnitude of the forces involved.

Thanks to a combination of math, physics and astronomy, we know quite a bit about the Big Bang – and even more about the "inflationary moment" that followed it. For example, in the case of the Big Bang, we are pretty sure it happened about 13.8 billion years ago.

In the case of the inflationary moment (only discovered in the last three decades), we know some interesting facts.

Here's a tidbit about that remarkable, infinitesimal fraction of a second (to be precise – one over ten to the 44th power of a second). The force required to stretch the universe from a single point (or "singularity") to the minimum ("Planck unit") of space-time is roughly one "googol" bigger than all the mass and energy that exists in the universe at present.

You want to know what is a googol? It's ten to the 100th power.

There is no serious scientist alive who doubts these estimates, which flow directly from equations of motion and relavity, coupled with measurements taken when the universe's early lights became visible, approximately 360,000 years after the Big Bang.

And there is no engineer in the world – or at least none that I have met – who doubts that all of that craftsmanship required an infinite or near-infinite outside force. It may not be God, but it sure walks and talks like God. And the beginning of life is just as startling.

On the Origin of Our Species

Trying to solve the mystery of the beginning of life, or the beginning of the universe, can consume a book all by itself. But this book is about *the beginning of human life.* As amazing as the Big Bang and biogenesis are, they are remote in time, not testable and not reproducible in the lab.

The beginning of human life is much closer to us in time and is observable to the extent that we can sense the final product every time we breathe, every time we think and every time we fall in love.

Not only that, but we are constantly observing the formation of new life that results from the coupling of an adult male and female. We can observe the negative effects of dysfunctional parenting and the positive effects of new life that is nourished with food, educated in science and arts, and nurtured with love.

Few of those occurrences in the life of our species, or for that matter in the lives of our ancestors, can be explained by evolutionary biology or its twin forces of the struggle to survive and the urge to reproduce.

And yet, in academia and in the popular literature, the research and scholarly publications that predominate are the ones that are carried out within that theoretical framework.

Well, that might be an exaggeration. The intellectual straitjacket that controls discussion and research on a strictly Darwinian foundation exists within biology and anthropology, much more than within psychology, sociology, economics and political science, all of which assume that human beings not only act differently from animals, but *should* act differently from animals.

Interestingly, the biologists themselves have recently modified their own Darwinian model in a significant way. Leaving aside for a moment the confusion that exists as to the progression from apes to men, there is now a whole new model of natural selection.

The gradualism of old, which is the hallmark of Darwinism, is about to disappear, replaced by a narrative that describes explosions of new forms of life, and catastrophic disappearance of old forms. Biologists are still convinced that species evolved spontaneously, but it is no longer Gospel to hold that it all happened gradually.

Where before the prevailing analogy was of a smooth slope upwards from elementary particles to complex organisms, there is now a more subtle, more descriptive and fancier name. We have gone from a nice, smooth slope to a staircase.

Punctuated Equilibrium

In reality, as we shall see, recent discoveries in animal evolution have revealed more of a staircase model, with landings that show very little slope interrupted by huge steps ("saltations") in which a veritable explosion of complexity takes place whose cause cannot be predicted or explained.

Evolutionary biologists call it "Punctuated Equilibrium." Never mind that the term is neither grammatical nor particularly logical. It saves them the embarrassment of saying that they simply don't know how or why the big jumps in complexity came about....

As we shall see, we know even less about the how or the why of the final big evolutionary jump – the one that led to our species.

From Pre-History to the Human Era

Making up stories of extraterrestials and parallel universes to fill in where science is unable to tread or unwilling to accept the probable infusion of forces from outside the system is one thing for pre-historic events. It works somewhat when the theories relate to things that happened very long ago, and were not recorded. It gets a little more difficult to concoct theories as we approach the modern era – particularly the era in which human beings appeared and started making visual records of what was happening.

History, when written by humans, may be exaggerated. It may be biased. But even in its most slanted versions, recorded history provides a track record of actual events. It is not totally mythical.

Beginning some time between 50,000 years ago and the end of the last ice age (10,000 years ago), the history of our species takes physical, verifiable shape.

It was the age of hominids, like the one we call "homo erectus" and the one we call "homo habilis." How they became "homo sapiens" is a big part of what this book is about.

Of Hominids and Pre-Humans

As puzzling as some of the prior "saltations" (unexplained jumps) are, such as inorganic matter to living organisms, the transition from apes to men is that much more intractable.

An article in *The New York Times* appearing a little after the turn of the century (February 26, 2002), admits that "agreement breaks down completely on the question of when, where and how...anatomically modern humans began to manifest creative and symbolic thinking."

The author, John Noble Wilford, begins the discussion by admitting that Europe has a monopoly on the archeological evidence of modern humans. He quotes Dr. Clive Gamble (Director of the Center for Archeology of Human Origins at the University of Southampton) for a proposition that has been rather evident to us laymen, i.e., that Europe, which he calls a "little peninsula," also "happens to have a large amount of spectacular archeology."

Eventually, Wilford gets to the crux of the matter, when he states:

> *The uncertainty and confusion over the origin of modern cultural behavior stems from what appears to be a great time lag between the point when the species first looked modern and when it acted modern.*

As we shall see, that is precisely the conundrum that faces modern scientists, particularly anthropologists, who seem stuck in a paradigm that refuses to acknowledge a handful of moments in natural history that elude a strictly material explanation – that beg for an explanation that emanates from outside the empirical laboratory. Unlike physicists, mathematicians and engineers, who wax poetic when describing the Big Bang or the inflationary moment immediately afterwards (when the universe formed itself into a tiny unit of space-time), the anthropologist is stuck in his laboratory and refuses to consider other dimensions, beyond the purely physical.

For him, it's all about conventional wisdom. There is no room for real analysis. Just follow the crowd and conclude that humans spontaneously (by random mutations) acquired the ability to think in abstract terms and to consider the long-term consequences of our actions.

Why not accept the idea that we are automatons, following the instinct to survive and procreate? It's all tidy and spontaneous and Darwin said so.

It also makes no sense whatsoever. Yet it fits the narrative that we are just a better educated ape.

My reader will see, in the chapters that follow, how many different dimensions separate us from the animals. It is not just the extraordinary intellect, or the remarkable interplay of emotions, as they clash with our more rational selves, or even the fact that we are dreamers and lovers.

In the end, we will explore yet another dimension in the human species. It is perhaps the most puzzling of all – the one that most clearly separates us from the species that preceded us.

It is a trait that occasionally leads a person to sacrifice personal comfort for that of another. It is akin to other forms of human love, but it is purer, and less explainable from the standpoint of sociobiology. It is an intangible, somewhat mystical, physically immeasurable trait possessed exclusively by our species.

But now let's begin at the beginning, which necessitates a survey of what empirical science says about our ancestors. Are we really direct descendants of chimpanzees?

Most modern scientists say that we are, even though the facts, viewed under the lens of 21st century empirical discoveries, increasingly belie that conclusion.

Let's see what the moderns say about our immediate, evolutionary ancestors.

CHAPTER II
THE THIRD CHIMPANZEE

*A zoologist from Outer Space would immediately classify us
as a third species of chimpanzee, along with the pygmy chimp
of Zaire and the common chimp of the rest of tropical Africa.*
Jared Diamond, *The Third Chimpanzee*

*M*uch has been written about the first man and woman, but little of
it, including the book by the celebrated anthropologist quoted above, has
solid, scientific support.

In reality, the human genealogical tree is, right now, in shambles. Up
until the end of the last century, scientists thought they had a good handle
on our ancestry – to the point that every high school biology classroom
shows a progression of short, powerful, bent-over apes, following slightly
taller and more erect apes, who ultimately link up with a fellow named
"Neanderthal."

The very hairy Neanderthal fellow, with enormous muscles and a
rather intimidating, furrowed brow, is pictured walking right behind the
fellow we call "Cro-Magnon." Mr. "Cro" looks a lot like us; he/she is thus
typically referred to as "homo sapiens" or even "homo sapiens sapiens" to
denote a fully cognitive species of primates.

That nice, theoretical progression of primates to humans has begun to
fall apart. Recent discoveries have shed doubt on what was supposed to be
the last and most important link between us and our supposed ancestors.

The Neanderthal fellow just doesn't pass the test of ancestry. Here's what Jared Diamond says about him:

> *Our large brain was surely prerequisite for the development of human language and consciousness. One might therefore expect the fossil record to show a close parallel between increased brain and sophistication of tools. In fact, the parallel is not at all close. This proves to be the greatest surprise and puzzle of human evolution. Stone tools remained very crude for hundreds of thousands of years after we had undergone most of our expansion of brain size. As recently as forty thousand years ago, Neanderthals had brains even larger than those of modern humans, yet their tools show no signs of innovativeness and art. Neanderthals were still just another species of big mammal.*

Neanderthal "Bashing"

Richard Klein and Blake Edgar (Stanford scientists of whom I shall say more later on) readily admit that Neanderthal tools are pretty rudimentary and represent little improvement over animal tools used by both prior apes and modern chimpanzees and bonobos. In an earlier work, I laugh at their effort to be politically correct when they caution that these observations should not be taken as "Neanderthal bashing," which - funny as it sounds - they define as "a kind of paleo-racism that all caring people should resist."

Diamond is more blunt (or should I say less "caring") when he concludes that "for most of the millions of years since our lineage diverged from that of the apes, we remained little more than glorified chimpanzees in how we made our living."

But wait a minute! Didn't Darwin tell us that humans evolved from the big apes by gradual mutations that happened in the effort to survive? Doesn't natural selection apply to the evolution of humans?

Darwin's seminal work on human ancestry was written in 1874. At the time, there was little understanding of micro-organisms like the cell – let alone the complex molecule we now call the DNA. At the macro level, there was only one skull that could be attributed to what later was considered to be our immediate ancestor: Neanderthal. Darwin never got to examine that one Neanderthal skull. And he was quite confused as to what might be the key correlation in favorable traits to a likely ancestor.

To grasp how utterly confused he was, I need to quote an entire paragraph from his 1874 book, *The Descent of Man and Selection in Relation to Sex:*

> *In relation to bodily size or strength, we do not know whether man is descended from some small species, like the chimpanzee, or from one as powerful as the gorilla; and therefore, we cannot say whether man has become larger and stronger, or smaller and weaker, than his ancestors. We should, however, bear in mind that an animal possessing great size, strength and ferocity and which, like the gorilla, could defend itself from all enemies, would not perhaps have become social; and this would most effectually have checked the acquirement of the higher mental qualities, such as sympathy and the love of his fellows. Hence it might have been an immense advantage to have sprung from some comparatively weak creature.*

Based on the above reasoning, Darwin would have us look towards small size in the hunt for our animal ancestor. Put another way, Darwin thought that perhaps we should search for the "nerdiest" of our animal predecessors.

Modern evolutionary biologists look quantitatively in the opposite direction: They search for the largest possible brain size, rather than small stature, as a sign of evolving intelligence. Until recently, that meant focusing on Neanderthal.

But what about the species that preceded the Neanderthal. Who exactly are the "we" to which Diamond refers when he says that for most of the millions of years since "we" diverged from the apes, "we" remained "little more than glorified chimpanzees"?

As my reader will note, there isn't much hard data to go on. Anthropologists acknowledge that there was a "Great Leap Forward" at some point, perhaps 50,000 years ago. We know the leap landed on modern man; but we don't rightly know who the leaper was.

We assume it must have been an ape – or more precisely a cousin of chimpanzees. And so we pick up the story on this side of the leap, and talk about the modern European fellow that we call "Cro-Magnon."

The First Humans

Anthropologists today prefer the term "EEMH," which stands for "European early modern humans" to that of Cro-Magnon, which really is just the name of a cave in France. Calling the first humans "Cro-Magnons" just won't do, because very similar fossils were found in caves all over Europe, and not just France.

From tools found in the same locations, and carbon-dated to about the same time, it appears that all of these fellows are closely related, among themselves, and also closely related to us.

How closely related? Well, very. In fact, they are clearly our direct ancestors; nothing separates us from them, as far as anyone can tell, except possibly a little bit of body mass. We might be a little more slender, though that is hard to tell, since soft tissue is not preserved during the thousands of years that separate us from Mr. Cro – excuse me, EEMH.

In this book, I will develop the argument, based on the most modern data and unbiased analysis thereof, that all early, modern humans are, in fact, Cro-Magnons who probably started out in the Middle East and ended up spending winters in the caves of Northern Europe.

Interestingly, our Cro-Magnon ancestors stopped drawing nice pictures in caves. Scientists are pretty sure that by about 10,000 years ago, when the last Ice Age ended, it became unnecessary to live for long periods inside a cave. Hence, no more cave drawings.

Another thing that surprises scientists is that there are no pictures, either in caves or otherwise, other than the ones in Europe. That's also where sophisticated tools are found.

Nevertheless, the great majority of anthropologists insist that our species descended from great apes who walked the earth a few million years ago. They are convinced that we are connected with African apes, whom they call "pre-human" or "hominids."

My reader can judge if that's a stretch.

Fossils found in Asia (for example, the so-called "Java Man") or in

Africa (e.g., the Australopithecus Afarensis whom we call "Lucy," because of the eponymous Beatles song) were definitely not modern humans. And they certainly were not European early modern humans (EEMH), as we now use the term.

As it happens, I had some personal connection to EEMH, since I was privileged to get access to what is perhaps the best known of the European caves where EEMH decided to draw startlingly accurate pictures of animals on the ceiling.

It is called Altamira and was discovered by an amateur anthropologist who happened to be the grandfather of the owner of Banco Santander in Spain.

Altamira and Lascaux

You may have heard the story of Marcelino Sanz de Sautola, whose little daughter, while accompanying him in his digging pursuits, crawled through a hole in the wall of a cave and found animal pictures. Marcelino was first skeptical, then diligent in following her lead, then surprised to see that there were, indeed, well-drawn pictures in that now-famous cave in Altamira.

Lucky for me, Marcelino's grandson, Don Emilio Botin (principal stockholder of Banco Santander) arranged a private showing of the cave for me and my wife. Nowadays, all you get to see is a facsimile of the real cave, which is only open for serious academicians.

When I visited Altamira, the best carbon-dating available calculated the age of the paintings as approximately 13,400 years "BME" (before the modern era). That, to me, seems reasonable, given that the last ice age did not end until about 10,000 years ago. Expecting modern humans to subsist much before that is just not realistic, even if they found shelter in caves. I should add that finding caves to shelter from the wind was not enough for survival: they also needed to have animal hides, to cover themselves, and fire to heat themselves – as well as to scare off their cousins, the Neanderthals. (The hides were probably cut with flint stones; those were probably made from obsidian, which is a volcanic rock whose flakes are naturally much sharper than modern razor blades...)

The story of early, modern humans is thus the story of fire and flint, without which humans like us could not possibly survive. It is also the story of faith, which allows humans to discard their prejudices and their fear and build bridges of communication and collaboration that lead to scientific discovery, which then leads to survival from disease, droughts, floods and other natural disasters.

Faith in the concept of human brotherhood and sisterhood is also the only known way to avoid constant, and deadly, internecine strife. Winston Churchill once said, of the human condition, that "under sufficient stress – starvation, terror, warlike passion, or even cold intellectual frenzy – the modern man we know so well will do the most terrible deeds…"

Indeed.

We have written records of our ancestors that allow us to interpret their deeds back about 5,000 – 6,000 years; before that time, we must interpret marks on bones, scratches and imprints done with rocks which we call flints and remains of what might have been human-made fires. What that history tells us, and logic confirms, is that only those ancestors who saw a divine spark in other humans actually became civilized.

And so it is not so much Jared Diamond's "guns, germs and steel" that distinguish our species and fuels its progress; it is "fire, flint and faith."

But that's not the way modern anthropologists see it. They are desperate to connect us to the Neanderthals.

Hohle Fels and the First German

It's amusing to see how hard anthropologists try to connect our ancestors to the Neanderthals. The problem is quantitative: Neanderthals are separated from humans by an estimated 27 mutations, which cannot have happened in the short time that separates the two species.

The two species are also separated by time – with Neanderthals becoming extinguished about 40,000 years ago and humans showing evident signs of intelligence perhaps no earlier than 14,000 years ago.

And yet anthropologists are convinced that *homo sapiens* was not only painting but sculpting much before that. And what is their exhibit? It's German; it's a sculpture; and it's called Hohle Fels.

Here's a passage from Wikipedia that describes the latest findings of science:

The remains of at least five distinct individuals were found at Hohle Fels. In 2016, researchers successfully extracted the DNA from three samples taken from the Magdalenian period found at Hohle Fels. The tests were performed on two femur fragments, HohleFels10 and HohleFels49, and a cranial fragment, HohleFels79. The two femur fragments possibly came from one individual. HohleFels10 and HohleFels49 were indirectly dated to around 16,000-14,260 BP BP, while HohleFels79 was directly dated to around 15,070-14,270 BP. All three samples were found to belong to mtDNA Haplogroup U8a. The Hohle Fels samples were found to be genetically closest to other ancient samples from the Magdalenian, showing closest genetic affinity to each other and for other samples taken from the Swabian Jura, such as Brillenhöhle, while also showing genetic affinity for another Magdalenian sample, taken from the Red Lady of El Mirón, as well as a sample from the Aurignacian, GoyetQ116-1, taken from Goyet Caves.

Let me cut to the chase. Actual human skeletons, when carbon-dated using the latest techniques, show that our oldest ancestors are about 14,000 years old. Because the technology used is based on half-lives of certain carbon atoms, if there is an error in the calculations, the age of the fossils might be half of the 14,000. If so, that coincides with the only complete fossils found frozen in the European Alps.

Is that a coincidence? You be the judge. We're not going to tackle that analysis here. For our purposes, we can assume that humans equipped with our brain capacity appeared no earlier than 50,000 years ago – although actual, artistic evidences of human creativity are probably no earlier than 15,000 years ago.

As to who were our direct ancestors who presumably bred the Cro-Magnons, that is a very good question....For that we need to go back to "Lucy" and her home in East Africa.

In The Footsteps of Eve

To summarize, there is not much that can be called "settled science" as to who were the first of our species. Evolutionary biologists agree only on two things: we came from sub-human apes like "Lucy," who made her home in East Africa about three million years ago, and we somehow made a "Great Leap Forward" from our ape-like ancestors some 50,000 years ago.

But, wait a minute. If we are direct descendants of Lucy, who lived in East Africa a couple of million years ago, how did we get from Lucy to whoever it was who drew the pictures in the caves of Altamira, Spain and Lascaux, France?

Good question.

Besides the problem of the Neanderthal not being our ancestor, there is a bigger problem: Lucy may not be in our evolutionary line at all. And that was millions of year ago.

The interesting tale is told by an eminent evolutionary biologist named Dr. Lee Berger. Tasked with analyzing fossils of pre-humans found in South Africa, Berger came up with conclusions that very much upset the anthropology establishment.

Up to that point, they were hoping that they could establish a very smooth transition in the evolution of the human species. Theoretically, we are supposed to have descended from the chimpanzees, yet as Jared Diamond readily admits, "almost no ape fossils of any sort have been found for the crucial relevant period between five and fourteen million years in Africa."

That's a huge gap; or as we used to say a huge "missing link" in the supposed chain from apes to men.

As strange as that missing link is, the next stage (from Lucy to us) is even more puzzling. Let's talk about Lucy and her great, great grandchildren; let's see if there is any kind of viable succession from the big apes of Africa to the cave-painting Cro-Magnons of Western Europe.

Let's review how that story went from a supposed clear progression to a lot of dead ends – how Lucy never made it to Eve.

From Lucy to Eve

In the nineteen seventies, a couple by the last name of Leakey (Mary and Louis) found some fossils that resembled our species in East South Africa. The skeletons were clearly those of bipedal (meaning two-footed) apes.

It was the highpoint of the Beatles musical group, and their songs were very, very popular. One of those songs was named "Lucy in the Sky with Diamonds," which many thought was code for LSD.

In any case, it was known popularly as "Lucy," and that became the common name for the species of apes who was proposed to be our lineal ancestor, going back perhaps three million years.

Lucy is more properly known as "Australopithecus Afarensis," which means little more than "old ape." Lucy and her family members had slightly bigger brain cavities than other great apes and used tools that had some similarity to those used by more advanced species, including our own.

Yet the possibility that Lucy and her cohorts were our direct ancestors is now pretty much discarded. One important reason is that their tools were very crude. Here's what Diamond says about that:

> *The only surviving tools from this period are stone tools that can charitably be described as very crude, in comparison with the beautiful polished stone tools made until recently by Polynesians, American Indians and other modern Stone Age peoples. Early stone tools vary in size and shape, and archaeologists used those differences to give the tools different names, such as 'hand axe,' 'chopper,' and 'cleaver.' These names conceal the fact that none of those had a sufficiently consistent or distinctive shape to suggest any specific function, as do the obvious needles and spear points made by the much later Cro-Magnons.*

Diamond expands on the difference by explaining that such stones, used by "homo erectus" were a far cry from the sophisticated tools of modern homo sapiens and concludes:

Many advances in tools that appear after the Great Leap Forward were unknown to 'Homo erectus, and early 'Homo sapiens.' There were no bone tools, no ropes to make fishnets and no fishhooks. All the early stone tools may have been held directly in the hand; they show no signs of having been mounted on other materials for increased leverage as we mount steel axe blades on wooden handles.

Diamond is clearly frustrated that there is no progression during the next few million years – despite the increase in brain size of these primates. Yet he is determined to believe (and have us accept) the idea that there was something important about to happen, and that it happened half a million years ago, instead of 50,000 years ago, after the Great Leap Forward.

Here's Diamond at his most puzzling, and puzzled, self:

Was our meteoric ascent to 'sapiens' status half-a-million years ago the brilliant climax of Earth history, when art and sophisticated technology finally burst upon our previously dull planet? Not at all: the appearance of 'Homo sapiens' was a nonevent. Cave paintings, bows and arrows still lay hundreds of thousands of years off into the future. Stone tools continued to be the crude ones that 'Homo erectus' had been making for nearly a million years.

And so it's safe to say that it was a "nonevent" all right, meaning something that never happened.

But, wait, we have not discussed the discovery of fire. Could that be the event that transitioned pre-humans to humans? Taming fire is certainly something that distinguishes our species from modern chimps and other primates.

As it turns out, that also seems to have been a more recent discovery than we might have thought.

We will explore that in a later section, when we tackle in earnest the two discoveries that made it possible for our ancestors to roam across the continents and reach the colder regions of the earth.

Of Peking Man

But what about "Peking Man" and other hominid fossils who show up in caves with tools and animal fossils that suggest they were adept at hunting and killing large mammals?

Here's what Diamond explains about the argument, made by many antrhopologists, that our human ancestors have been "successful big-game hunters for a long time":

> *The supposed evidence comes from three archaological sites occupied around 500,000 years ago: a cave at Zhoukoudian near Beijing, containing bones and tools of 'Homo erectus' ('Peking Man') and bones of many animals; and two non-cave (open-air) sites at Torralba and Ambrona in Spain, with stone tools plus bones of elephants and other large animals. It's usually assumed that the people who left the tools killed the animals, brought their carcasses to the site, and ate them there. But all three sites have bones and fecal remains of hyenas, which could equally well have been the hunters.*

Based on all the evidence, Diamond concludes that nothing important happened between Lucy and the Great Leap Forward. He happily (and humourously) concludes that it's all a "mystique" created by alpha-males to connect us to our big-game hunting ancestors.

Male anthropologists, he argues, have been "trapped in this mystique" to the point that they have engaged in "locker-room mentality" and in the "purple prose" that surrounds this, which he calls "pure fantasy."

By way of illustration, he cites a quote that I myself found phantasmagorical enough to cite in a previous work, by Robert Ardrey in his very readable (though not very scientific) book called *African Genesis:*

> *In some scrawny troop of beleaguered not-yet-men on some scrawny forgotten plane, a radian particle from an unknown source fractured a never-to-be-forgotten gene and a primate carnivore was born. For better or for worse, for tragedy or for triumph, for ultimate glory or ultimate damnation, intelligence made alliance with the way of the killer, and*

Cain, with his sticks and his stones and his quickly running feet emerged on the high savannah.

It is, indeed, fantasy. And many of us, using the best evidence available, actually think that early humans were very possibly vegetarians. Of course, you can't live off the land if the land is frozen, as it was in many places before the end of the last Ice Age, roughly ten thousand years ago.

In order to survive in the frozen caves of Northern Europe, as our ancestors clearly did, you had to have fire and flint, and be able to scare away, if not kill, the bigger mammals.

And who would have been our competition? Well, just about any large predator that did not possess enough intelligence to wield fire and tip spears with sharp flint-stones that would wound a bigger, more powerful animal.

The fossil record had, until recently, been science's main source of information; nowadays it is easier to analyze it using genetics. It is in that discussion that the genetic differences in species comes in handy.

Clues Offered by DNA

The discovery of modern genetics, particularly the molecular chains we call DNA, for deoxyribonucleicacid, should by all rights give us fail-safe technical tool for comparing species.

Alas, it is not that easy. For one thing, species that seem quite different in functionality, size and intelligence are often quite close genetically. The opposite is also true: Species that seem far apart genetically often have common traits and appear quite different on the outside, yet have interchangeable internal organs.

A good example of the latter, as far as being similar to our own species in functionality but not in DNA, are the guinea pigs. In skin care matters the guinea pig is quite similar in traits to the human organism.

To an amateur like myself, and probably to my reader, the entire analysis based on DNA is likely to be unconvincing, for at least two other reasons.

The first reason is akin to the one that debunks brain size as an indicator of intelligence. As mentioned before, the Neanderthals had bigger brains than many humans, but they were about as dumb as any of the big apes.

Using the computer analogy, DNA is about hardware; intelligence is about software. Diamond himself clues us in on the oddities of genetic similarities when he informs us that "our genetic distance from chimps" is 1.6 percent, which is less than "half the distance of orangutans from chimps (3.6 percent.)"

So by this measure we are more "chimp-like" than orangutans!

I kept on reading this chapter of the celebrated Diamond book to see if there was a technical explanation of how we calculate "genetic distance." Could the percentages be calculated by how many genes we have, or how many chromosomes, or how similar the genes are in molecular formation, or some other quantitatively significant measure?

Well, get ready for this. The way those "genetic distances" are measured is based on a discovery by molecular biologists that relates the longevity (time difference) that is thought to propel changes in a DNA. Rather than try to explain it myself, let me simply quote directly from Diamond:

> *A quick method of measuring changes in DNA structure is to mix the DNA in two species, then measure by how many degrees of temperature the melting point of the hybrid (DNA) is reduced below the melting point of pure DNA from a single species. The method is generally referred to as DNA hybridization. As it turns out, a melting point lowered by one degree centigrade...means that the DNAs of the two species differed by roughly 1 percent.*

I had to think very hard about that one. Perhaps being a mechanical engineer and not a chemical engineer, I didn't grasp the chemistry inherent in this method of comparing strings of molecules made up of a couple of billion similar molecules.

Then I realized that the *percentage* measuring stick being used was based on the fact that from freezing water to boiling it, we count one hundred separate, equal gradations. Water freezes at zero degrees centigrade and boils at 100 degrees centigrade. That's where the word "centigrade" comes from!

I was dumbfounded by the easy way in which anthropologists used chemistry – or more properly, biology mixed with chemistry – to determine genetic differences.

It really doesn't matter, because even if this particular measure of genetic difference made any sense, all it measures is *hardware* and what counts for intelligence, and psyche and whatever it is that makes our species able to change our minds, and apologize and love, or hate, is either our *software* (referred to by modern scientists as our "mind-psyche") or a very complex combination of our mind-psyche (software) and some little voice inside of us that constitutes our conscience and which illuminates our will.

Diamond and his cohorts in sociobiology have done their best to convince us that the physical differences between our animal ancetors and us is really just quantitative and not qualitative.

My reader can judge from the chapter that follows, which delves into what now appears as a quantum leap in the progression from apes to humans.

By their own admission, it was an unexplainable "great leap forward."

CHAPTER III
THE GREAT LEAP FORWARD

Then something mysterious happened: Jared Diamond calls it 'the great leap forward'; anthropologist Marvin Harris calls it 'cultural takeoff.' Whatever caused it, its results were soon apparent: with the aid of a greatly expanded technology our species spread out over Europe and Asia, and right about that time the Neanderthals ceased to exist. Judith Rich Harris

*J*ared Diamond is a Pulitzer Prize recipient; his bestseller, "Guns, Germs and Steel" is the most quoted, most current and most revered treatise on anthropology.

By all rights, anthropology should be a "hard" science. After all, the subjects to be examined are easy to identify and examine. In fact, the anthropologist, of all scientists, is the only one who can analyze his subject by both introspection and by empirical testing, as is done in a biopsy or an autopsy.

Using the terminology of philosophy, anthropologists, like psychologists, are able to investigate by using both "inductive" (measurable experiments) and "deductive" methods (by which is meant basically analyzing our own internal thoughts and making logical connections).

However, for reasons that I cannot discern, the typical anthropologist refuses to focus on the introspective "discovery horizon." I use the term "discovery horizon" as it is used in astrophysics; when trying to discern what happens inside a black hole, which does not emit light waves, the astrophysicist peers inside the black hole by extrapolating from what happens right at the "horizon" of what is observable and measurable.

In a similar vein, the anthropologist, like the psychologist, can peer inside the human psyche by looking instrospectively inside his own psyche – his own black hole, where no one else can penetrate. Yet anthropologists seem programmed to ignore any evidence that comes from within their own psyches.

Instead, they make an *a priori* decision that man is just another material object, which happens to have classic animal hardware and a very sophisticated, built-in software, but nothing more. Man, to them, is the visible part of the universe, but not the invisible part, including the black holes (which, parenthetically, are estimated to constitute about a quarter of all matter in the universe). By not peering into the invisible part of the human reality, or making logical deductions from the empirical data (tests, measurements, statistical data), anthropologists miss at least one-third of the human equation.

If astrophycists did that, we would not know about the Big Bang or the inflationary moment immediately after the Big Bang, when the universe took shape in a tiny fraction of a second. Those discoveries require mathematical extrapolation, based on the logic of equations, but not on actual data. (That's why the science is called "theoretical physics.")

The same kind of extrapolation, or theorizing from actual data, is needed to discover and understand the periods of time after the Big Bang. Without such theoretical calculations, we would know nothing at all about the universe for the first 360,000 years after the Big Bang (the so-called "Deep Field"), when light was not yet a viable phenomenon in nature.

The social sciences use similar, theoretical methodology to peer inside the human personality. Their analysis begins with empirical data about the human body, including the brain; that data is easily comparable to simple biological realities, in which humans and animals are quite similar.

But the analysis doesn't end with empirical realities. Most social scientists (philosophers, psychologists, ethicists, theologians) allow themselves to delve into more introspective realities, many of which can only be measured indirectly (analogous to the astrophysicist's "event horizon").

We see such insights playing out among philosophers. In that field of learning, one notices that the language of science moves away from *man-as-smart-animal*.

It becomes more like *man-as-different-animal*.

For example, in a recent *New York Times* article, Clemson philosophy professor Todd May describes humans as having "an advanced level of reason that can experience wonder at the world in a way that is foreign to most if not all other animals."

He gets more specific in his description of our distinct, analytical powers, when he says:

> *We create art of various kinds: literature, music and painting among them. We engage in sciences that seek to understand the universe and our place in it....*

Diamond uses a similar string of attributes to distinguish humankind from other animals when he refers to the distinctly human attributes of "innovation, art and complex tools." I rather like that troika of attributes and will pretty much adopt it for the ensuing discussion.

Note, however, that innovation is not as easy to pin down as art and tools. Innovation implies creativity; and creativity implies original thoughts.

These attributes are not particularly compatible with evolutionary biology. And therein lies much of the confusion in the scientific discussion of how hominids became humans.

The best that science can do is refer to the final jump, from erect, bipedal, big-brained primates to full-fledged homo sapiens, as "the Great Leap Forward."

And what a leap it was.

The Great Leap

The last jump ("saltation" in modern, scientific lingo) is described by Diamond as happening, "at least in Europe," when the mentioned attributes (innovation, art and tools) "appeared unexpectedly suddenly, at the time of the replacement of Neanderthals by Cro-Magnons."

Anthropologists constantly make such statements, yet they fail to parse them, to see where they lead us, in terms of chronology and causality.

In a later section, I will analyze the timing and causality of the discovery of fire, which constitutes the first quantum leap in the supposed chain of intellectual transformations from humanoids to humans. My analysis will be based on facts, and on the logical connections between those facts and the use of fire. I will often use analogies and comparisons based on engineering.

In other words, my analysis will be based on hard science.

Ostensibly, that is the same methodology used by Jared Diamond in the book from which I derive the title to this chapter. Yet Diamond is hampered by a presupposition that hurts his analysis: He assumes that the evolution of humans from animals was due to natural selection (survival of the fittest). For that reason, his analysis is mostly strained, if not outright fanciful.

Let me illustrate with a couple of examples.

According to Diamond, "natural selection surely explains *some* geographic variations in humans." He offers as an example the fact that "many African blacks but no Swedes have the sickle-cell hemoglobin gene, because the gene protects against malaria, a tropical disease that would otherwise kill many Africans."

Okay, we can buy the logic, although there are plenty of other plausible explanations, including the most modern of all, which goes by the name of "epigenetics" and means, simply, that some acquired traits are passed on to the next generation. Epigenetics would suggest that Africans avoided contracting malaria by some adaptive mutation(s). Such adaptive mutations were then passed on to the next generation.

But it is also possible that Africans simply used their inventiveness to find a cure, which they then shared it with others of their tribe or nation. Indigenous people, parenthetically, have found many native remedies that are still in use in modern medical science. By way of example, I am general counsel to a pharmaceutical company that uses, as active ingredients, urea (urine) and papaya for wound-care of patients. Both of these cures date back decades, if not hundreds of years.

They were classic, indigenous medical inventions – hardly the result of undirected micro-mutations resulting in natural selection of the fittest.

About Skin Color and Evolution in Humans

The classic case that most of us believe shows how humans evolved in the effort to survive is the one dealing with skin color. Alas, it is not that simple.

Let's discuss that one.

Here's what Diamond himself says about that:

> *Our skins run the spectrum of black, brown, copper, and yellowish to pink with or without freckles. The usual story to explain this variation by natural selection goes as follows. People from sunny Africa have blackish skins. So (supposedly) do people from other sunny places, like southern India and New Guinea. Skins are said to get paler as one moves north or south from the equator until one reaches northern Europe, with the palest skins of all. It's...obvious what good a dark skin does in sunny areas; it protects against sunburn and skin cancer.*

Sounds entirely logical – what commentators like to call "settled science." But it probably has no validity whatsoever, as Diamond readily admits:

> *Unfortunately, it is not so simple at all. To begin with, skin cancer and sunburn cause little debilitation and few deaths. As agents of natural selection, they can have utterly trivial impact compared to infectious diseases of childhood.*

Diamond documents all kinds of examples of people who are white and blond and yet inhabit sunny environments. And then there are the American Indians, none of whom have dark skins, even in the sunniest parts of the hemisphere.

As many as eight other theories are explored; none sound particularly convincing. And so Diamond admits that Darwin himself despaired "of imputing human racial variation to his own concept of natural selection" and quotes him as saying that "not one of the external differences between the races of man are of any direct or special service to him."

29

If my reader is surprised that race does not fit into evolutionary theory, wait until you hear how difficult it is to apply its principles to more important human traits, such as those that determine how we age.

Why and How Humans Grow Old

The human equation on aging is quite complex, as we shall see. But that does not deter the evolutionary biologists, who are desperate to confirm that all human traits respond to the instinct to survive long enough to reproduce.

Right off the bat, the first obstacle in that effort is that women are only fertile during perhaps a fourth of their expected lives. Says Diamond:

> *Most mammals, including human males plus chimps and gorillas of both sexes, merely experience a gradual decline and eventual cessation of fertility with age, rather than the abrupt shutdown of woman's fertility. Why did that peculiar, seemingly counterproductive feature of ours evolve?*

Diamond really can't explain it. He also doesn't give satisfactory answers to the question of why our species ages so slowly. But he sure tries. Here's the gist of his syllogism:

> *Slow aging is...crucial to the human life-style....That's because our life-style depends on transmitted information. As language evolved, far more information became available to us to pass on than previously. Until the invention of writing, old people acted as the repositories of that transmitted information and experience, just as they continue to do in tribal societies today. Under hunter-gatherer conditions, the knowledge possessed by even one person over the age of seventy could spell the difference between survival and starvation for a whole clan. Our long life span, therefore, was important for our rise from animal to human status.*

All right. So our species needs to grow old because we rely for our survival on skills passed from one generation to the other – as opposed to the animals who pass those skills very early in life, even though they cannot talk or write.

Perhaps now that we have the internet and oodles of libraries with oodles of information on how to survive, our bodies will atrophy and cause us to die at an early age.

But that's not happening; instead, humans are living longer and longer.

That's just another dead end in the current scientific literature, which seeks to explain everything in terms of Darwinian natural selection.

But what about the discovery of fire? Isn't that the most compelling marker in the evolution of our species? And what about the increased sophistication in the tools made by our pre-human ancestors? Don't those show a clear progression from East Africa's "Lucy" to our cave-dwelling, European ancestors?

If only it were that simple.

The Discovery of Fire

Humans, in order to escape what must have been a very restricted homeland, needed an external source of energy. Even if we assume that the first homeland was in the Middle East, any venture north of the Equator would involve a drastic lowering of temperature and a corresponding need to have some sort of heating element.

From every recent archeological discovery, it is easy to conclude that our ancestors first ventured north to Germany, France and Spain around the time of the pictures found in caves in Western Europe. To my engineering mind, that could not have happened before 15,000 years BME.

Fortuitously, the *best* evidence for these excursions is really the *only* evidence we have of the earliest man and woman. While the date is not clear, the fact that humans were able to claim the subterranean real estate (which we call caves) indicates, with a fair degree of scientific certainty, that they had the use of fire. Some of the caves were in France and from there we derived the term "Cro-Magnon;" it was sometime between 15,000 and 30,000 years ago.

I am not going to quibble on the exact date. I am fairly sure that it could not have happened earlier than 14,000 BME; but was it much earlier, as many scientists would now argue?

(In another book by this author, I relate the story of delving into an enormous tome on ancient art, which my godfather had in his ample, Key Biscayne home. As I looked at photograph of Lascaux and Altamira cave drawings, they were all carbon-dated to between 13,000 and 14,000 years ago. But then, to my surprise, each of the dates given had an asterisk next to it, which referred the reader to the bottom of the page, where it was pithily explained that "scientists have recently doubled all the estimates…")

Evidently, there has been an effort to push back the advent of humankind, as well as the related discovery of fire. But, as previously discussed, there is now scientific consensus that Neanderthals did not have the use of fire. So it begs the question of when and by whom was fire discovered.

Let's discuss that.

Dating the Discovery of Fire

It should be pointed out that fire is not a human invention. It's not even a human "discovery." Fire has been with us ever since the Big Bang.

The question is, *when did humans learn to manage fire?*

And that is a tough one. How does one separate a spontaneous fire, which humans may have used to cook food or forge a better weapon, from the ability to start a fire intentionally.

I turn to Stanford anthropologist Richard Klein, who is not only the most recent to prounounce himself on the discovery of fire, but also the most honest. Klein acknowledges that he has done extensive research and is unable to pinpoint the who, when and where of the first little fireplace built by our ancestors. Having reviewed all sources available to a twentieth century Stanford University researcher, he and his co-author (Blake Edgar) "lay bare their own prejudice" and go with the date of 250,000 years before the modern era.

The problem with that date is that it is a "when" that is unaccompanied by a "where" or a "who." Let's see if we can find a "who."

From Neanderthals to Humans in the Caves of Europe

We are now pretty sure that Neanderthals were kicked out of caves because they didn't have the use of fire. That was about 40,000 years ago. But early humans, according to Klein, developed the amazing capacity to think in abstract terms about ten millennia earlier than that.

As Klein and Edgar tell it, humans evolved in no less than four quantum leaps; the last one, which is clearly the most dramatic one, took place around 50,000 years ago and entailed a "genetic mutation that promoted the fully modern brain." This quantum leap, which they admit "cannot be tested...by experiment or by examination of relevant human fossils," enabled humans to "adapt to environment not through anatomy or physiology, but through culture."

The cognitive *coup de grace* of this remarkable and unexplainable mutation, according to these illustrious modern scientists, was that humans acquired a trait that allowed us to "conceive and model complex natural and social circumstances entirely within our minds." This, in turn, equipped us with "the full-blown capacity for culture, based on an almost infinite ability to innovate."

Wow! In the span of no more than 10,000 years (about the same span of time as recorded human history), humans passed their first cousins, the Neanderthals, who were deathly afraid of fire (as animals typically are) and went on to have "an almost infinite ability to innovate."

It actually may have happened a lot later than 50,000 years ago; but we don't have time to quibble. All we know is that by the time of Altamira and Lascaux, our cave-dwelling ancestors had learned to paint pictures of animals that are more accurate than anything I could draw.

By that time, they also made some sophisticated tools, beginning with the most important of all – the cutting edge, aka the "flint" made famous by the television show called "The Flintstones."

But let's backtrack a bit and consider some of the more elementary tools made by pre-humans. Here we will discover the biggest misconception of ancient, pre-human history.

Book after book, essay after essay, interview after interview, all commentators (including scholars) convey the impression that there is a neat progression in tool-making among our animal predecessors.

Let's see if that progression holds.

Tool-making in the Animal Kingdom

One would think that tool-making is the best measure of intelligence in the animal world. But it is a spectacular fallacy to conclude that there was a gradual progression, over time, that ended with primates, who look the most like humans. I suspect my reader will be surprised to learn that animals which look the most like us (including the big apes classified as "hominids") were not necessarily the most ingenious tool-makers in the animal kingdom. Hominids are not even among the top three, as we shall see.

If intelligence is analyzed in terms of the sophistication of tools that animals make, then the ranking of smartest animals might have to look somewhat like this.

First Level: Insects.

What ants and termites do is just beyond belief, given the small size of their brains. A recent article by the Associated Press covered the discovery of termite mounds in northeastern Brazil; these were built by a species called "syntermes dirus" and they cover an estimated 230,000 square kilometers, which is an area roughly the size of Great Britain.

This book is not precisely about zoology, or engineering for that matter; but it's worth mentioning that these little critters built approximately 200 million "cone-shaped soil mounds that are 2.5 meters tall and approximately 9 meters in diameter." The AP article follows that description by clarifying that "these mounds are not nests, but rather they are generated by the excavation of vast inter-connecting tunnel networks."

So why did these little insects build these tunnels? Well, it's simple: to "allow them to access dead leaves to eat safely and directly from the forest floor."

Heck, we should have known that; it is really no different to what mole rats do, "which also live in arid regions and construct very extensive burrow networks to obtain food."

Wait a minute. I said that mole rats do that. But I meant to say "naked mole rats." Goes to show that hiding your nakedness is not a particularly high priority for rodents. That brings us to....

Second Level: Beavers

As an engineer, I am always taken by beavers, for the simple reason that they build dams. Are you serious? A small little rodent taught *homo sapiens* how to control the flow of significant rivers?

Yes, Ma'm. That is correct. Beavers build entire structures that channel or simply stop the flow of large bodies of water.

O.K, but birds build nests on trees and light poles and whatnot. And spiders build complicated webs strung together from filaments that they produce while in mid-air.

Those high-flying structures are complex, from the standpoint of structural engineering. The thing about beavers is that they not only build dams, but also hide green leaves and live branches under the muddy bottom of rivers, so that they can retrieve them when the rest of the world freezes and there is no plant alive to be devoured.

They are my no. 2 species in the realm of smarts.

Third Level: The Birds

At one point, scientists were sure that birds evolved from a little dinosaur who got tired of climbing trees to avoid bigger predators, and flapped his limbs frantically until, over time, the little creature developed wings.

It was called the *archaeopteryx.*

I had a client whose hobby it was to identify new species of archaopteryx. He did this by exploring and exporting back to his place in New York rocks containing fossils of this little animal that were found in some remote part of China. Once in New York, he would classify the little fossils, describe the differences from prior species he had found, and submit them to the local science museum as new species.

The local museum, enthralled that he had done all this work – and that the museum would get to classify some new species – would name them after his daughters. (It was eerily reminiscent of the way that Galileo managed to get the moons of Jupiter named after his benefactor, Cosimo de Medici.)

It turns out that the little dinosaur, archaeopteryx, is not an ancestor of the *aves* family, of which the modern birds are part. Like most other major animal branches, we do not know what preceded them. Nor do they seem to have resulted from slow, gradual, evolutionary changes; in fact, most major animal branches were produced in a relatively short period of time called the Cambrian period (about 500 million years ago).

Anyhow, back to the birds. Besides the amazing nests that birds construct and which seem to resist all but the fastest winds, birds are amazingly wise. They are also versatile: they can walk, fly and swim.

In the Galapagos, where Darwin did most of his research, the birds range from high-flying finches to the rather stodgy cormorant, which cannot fly at all. (It is not hard for a flightless bird to find food in the Galapagos; two currents that meet there, the Humboldt and Cromwell, bring every kind of fish and anphibian, almost as if it were a five-star hotel, where cooks and waiters bring the morsels right to your seaside table.)

And it should be noted that we humans have adopted not only the structure of the limb that allows birds to fly, but even the name: Our airplanes are equipped with "wings" that allow objects that are much heavier than air to climb into the skies.

The simple notion that animal intelligence evolved by gradual increments from bacteria to insects to fish and reptiles, then to birds and mammals, ending with the primates and us, is simply not warranted.

Let's do a quick survey of some other species that have shown a high I.Q.

Of Black Bears, Whales and Other Smarties

One of the most recent and most illuminating studies of animal intelligence was published in the *Sierra* Magazine and written by Brandon Keim. The research, done on black bears no less, was the work of an amateur lover of wildlife named Ben Kilham.

> *Starting with two little cubs who were abandoned during a logging operation that scared away their mother, Kilham followed the growth to adulthood of one cub, named "Squirty," who for unknown reasons remained close to Kilham's home for the ensuing two decades.*

Here's a sample of Kilham's observations and findings:

> *Bears are quite social...they have a society of sorts, a matriarchy that in this case is governed by Squirty; they have a rich system of communication; they are highly self-aware; and perhaps most surprising, they are governed by long-term relationships and rules of conduct.*

As is probably the case with my reader, I have gotten used to such descriptions of animal societies and communicability – though not so much for black bears.

Scholars have done extensive research across species (and across rather distant chronologies covered by the first appearance of those species). Here's a sample of the human-like traits found in animals, as catalogued in *Google Scholar* and cited by Brandon Keim, who refers to the compilation as a "menagerie" of intelligence:

> *Ravens can plan for the future and demonstrate a degree of self-control comparable to great apes'. Sperm whales engage in consensus-based decision-making during the course of their travels. Japanese great tits, songbirds related to chikadees, use syntax – a linguistic property long thought unique to human language – when they communicate. Experiments show that tiny zebra fish, a species used to model basic animal traits, possess detailed memory of events and can learn from one another. Many species possess emotions. Giraffes appear to grieve, bumblebees show signs of happiness, and crayfish can experience anxiety.*

For every reason imaginable, the notion that animals improved gradually, whether by natural selection or otherwise, their level of intelligence, just doesn't seem to fit the chronological evidence. Intelligence in tool-making or communication does not increase over time from insects (400 million years ago) to dinosaurs (85 million years ago) to birds (60 million years ago) to great apes (5 million years ago).

Moreover, as mentioned before, it's interesting that almost all extant species began to appear in the fossil record around 500 million years ago, in the previously mentioned "Cambrian" period.

And almost all seem capable of doing the kinds of things catalogued in the cited *Sierra* Magazine. The fact that all of those animals, from birds to giraffes and bears, do not particularly look like us, does not mean much, as far as intelligence.

Let's review the evidence, in light of modern technology, of which computers are the prime exhibit.

Computer Hardware and Software

The best way to compare how animals and humans differ in how they process information is in terms of computer terminology. We use terms like "hardware" and "software" to distinguish the guts or body of a computer, which are built-in, from the various programs which are superimposed to provide more alternative "applications."

Using those terms, animal intelligence can be compared to computer hardware; and it can be as sophisticated, or more sophisticated, than human intelligence.

In other words, animals have amazing intuition. But that is a far cry from the ability to learn – or, in computer terms – acquire software from the surroundings, or from others of your species.

That brings us to a classic human trait called "collective learning."

Collective Learning

David Christian is a British scientist with a flair for words. He managed to condense a history of the universe into a marvelous 18-minute lecture packaged as one of the most popular "Ted Talks."

About two-thirds of the voyage from the Big Bang to the present, molecules combined into little one-cell organisms, and began inhabiting the earth we live in. Almost ninety percent of the way from those little organisms to us, a whole new set of complex organisms came into being. From the standpoint of information technology, living organisms brought to life a whole new form of software, which allows the accumulation of information.

But, as Christian explains in his Ted Talk, the information accumulated in the mind of an animal dies with his death. Humans, on the other hand, are able to share accumulated knowledge from generation to generation.

Humans are like a computer that can program itself. Our species uses its imagination, borrows external inputs from nature (including other creatures) and ultimately discovers and invents the kinds of artificial-intelligece (AI) systems that animals cannot even begin to grasp, let alone invent.

David Christian calls that trait "collective learning."

Collective learning begins in earnest with the invention of spoken language and reaches a fascinating climax with the invention of written language. As in the famous dictum by Rene Descartes, we humans are distinguished from the animals because we think ("cogito ergo sum"). But what good is thinking if we cannot convey our thoughts to others, or combine our thoughts with others to invent entire theoretical constructs?

What good is thinking if we cannot write down history for posterity, or reduce our theoretical formulas to equations, or sketch an architectural marvel or an artistic inspiration?

The ability of our species to communicate not only *within* generations but *across* generations is what distinguishes us from our predecessors. Like the political commentators on television, whom we see only from the torso to the head, we are "talking heads." And much of what we talk about is eventually transcribed into writings, including newspapers, magazines and books.

We call it literature and much of it is whimsical. Humans like to communicate with pictures, carvings, statues that convey ideas without using words. For some unexplained reason, anthropologists find it puzzling that our animal ancestors do not engage much in artistic pursuits.

Let's review some of their pronouncements on that issue.

Of Arts and Literature

Despite insisting that there was a very gradual evolution from chimps to humans, any honest scientist is heard to wonder, as Jared Diamond does, why no art is found "for the first 6,960,000 years since we diverged from chimps."

In reality, as exciting as paintings on caves are, art is not the main trait that humans have that distinguishes us from our supposed first cousins. Art is but a prelude to the really amazing new trait – the use of language.

As Diamond acknowledges, "human language origins constitute the most important mystery in understanding how we became uniquely human." And for those who would argue that we evolved naturally from animals, he says, rather forcefully, that "between human language and the vocalization of any animal lies a seemingly unbridgeable gulf."

Indeed.

Diamond explains that the most sophisticated "animal language" so far discovered is "that of a common cat-sized African monkey known as the vervet." He elaborates by giving the example of the vervets of Amboseli, who seem to possess at least ten "putative" words, to wit: leopard, eagle, snake, baboon, "other predatory mammal," "unfamiliar human," "dominant monkey," "subordinate monkey," "watch other monkey," and "see rival troop."

That's exciting, though I have also read that elephants emit as many as 80 distinct sounds. Compare that to 142,000 words for a typical English dictionary or to the 65,000 – 80,000 that Winston Churchill used in his voluminous writings.

But even that comparison does not suffice to explain the difference in ability to communicate. Humans don't just memorize tens of thousands of different words as they grow up; humans combine them into sentences and pararagraps that have varying sequences and combinations, which convey entirely different ideas.

Churchill wrote about 10 million words, combining them such that they offer us almost unlimited ranges of thought.

The words of each sentence I write in this book can be recombined to say something entirely different – or to make no sense at all. My mind combines them by using a form of processing that is impossible even for a very complex computer.

It's been estimated that creating a computer already hardwired to do the mental tasks of the human mind would require, using the most modern micro-chips, a machine that would have the size of a galaxy.

In other words, impossible.

Thinking that it happened by pure accident, as our ancestral chimps worked hard to survive, is a fable as hard to believe as those which we see in today's science-fiction movies.

So now, let's delve a little into the human brain. This organism has only recently been the object of empirical analysis. It's not pretty or neat. But it sure is interesting.

CHAPTER IV
YOUR BRAIN, EXPLAINED

Our understanding of the brain is always evolving and – like other scientific disciplines – neuroscience is constantly correcting itself....Although we can't ever be certain the conclusions we make about the brain are representative of reality, we can be confident any mistakes we've made will eventually be rectified by later scientific investigations.
Marc Dingman, PhD

*I*t is daunting to tackle the study of the human brain. One reason is that the human brain is an extraordinarily complicated machine. God bless the neuroscientists among us, for they do this kind of analysis on a daily basis.

There is a common expression, when emphasizing that something is not that difficult to understand, which proclaims: "It's not rocket science."

Rocket science, as complicated as it is with about a million or more moving parts, is child's play compared to neuroscience.

Here's why.

In a marvelous new book, titled *Your Brain, Explained*, Penn State professor Marc Dingman, Ph. D., describes the human brain by comparing it to that of a sea slug (more technically known as *Aplysia Californica*). Both of the brains in question are composed of many little signal-carrying characters known as "neurons." These little codgers connect with one another by bearing their signal flags to the edge of tiny precipice called a "synaptic cleft," where they are picked up by fellow neurons on the other side of the tiny crevice.

How tiny is the clever, synaptic cleft? Funny you should ask. Try 20-40 nanometers. And how little is a nanometer? Funny that you ask that too; by comparison, a human hair is about 100,000 nanometers wide!

Neurons are as numerous as they are tiny. If we go back to the comparison to the friendly sea slug, here's what we find. Whereas the slug brain has 20,000 neurons, the human brain has 86 billion. And, to complete the comparison with an in-between critter, a mouse brain is equipped with 75 million neurons....

So, to conclude, one difficulty with understanding the human brain is the sheer complexity of it – the remarkable magnitude of its tiny signal-senders and receptors.

But the other reason is perhaps more intractable. Unlike human-made machines, in which the component parts play unique roles, the human brain is not so easily compartamentalized. For example, even though neuranatomists have identified our ability to see as emanating mainly from the "occipital" lobe (in the back of the head), the issue is not so simply solved. Explains Dingman:

> *Studies suggest...that there are over thirty areas of the cortex involved in vision and only about one-third of them are found in the occipital lobe. Indeed, most of the functions of our brain seem to be distributed through the organ, rather than concentrated in one area.*

And so it is, but not only for our brains. It is that way for our entire human person. We are a combination of mind-psyche, body parts, ideas and inspirations that are hard to pin down when trying to determine what drives our actions.

And the toughest to pin down is our will to love and to share, even under the most difficult circumstances of deprivation. Interestingly, and perhaps for that reason, the most popular, imagined center of the human personality – as far as its flights of nobility – is pictured as the human heart.

I find that very interesting, because most of the processing of information is done in the brain. Yet, instinctively, we know it's a complex combination of thoughts, feelings, ideas and even external motivators that act upon our will to make us fully human.

We are not what we eat, or even what we think. *We are a combination of our needs, our aspirations and our loves.*

Human Decision-Making

Undoubtedly, the human brain is unique. It resembles an animal brain, but only up to a point. Unlike an animal brain, which is like a laser beam that focuses on survival, the human brain is more of a tug-of-war between emotions and rational constraints.

Dingman, in his very recent (2019) and very readable treatise on the human brain, describes the battle for control that characterizes human decision-making – i.e., the battle between emotions and reason. One part of the brain (the limbic node) is linked to what he calls "primitive forms of behavior," i.e., "the types of pleasure-seeking, pain-avoiding stuff that drives many of our actions."

Dingman proceeds to explain the balancing act that takes place when other, more rational part of the brain compete with those "primitive forms of behavior," which –

> *we like to think we can suppress with the more rational parts of the brain. This type of impulse control is not as commonly seen in other animals and consequently some consider it a feature that helps separate us from other species.*

Evidently, we humans are not always able to superimpose our rational side on our more primitive forms of behavior; often we let our "pleasure-seeking/pain avoiding" dimension get the better of us. "Negative" (destructive) impulses often rule the day and the consequences can be quite damaging to ourselves or to others.

As mentioned before, the tug-of-war that takes place within the human psyche is best understood as a three-dimensional battle for domination. As with so many other things in nature, a digital or binary explanation will not do.

Moreover, because the battle is not purely physical (i.e., the contestants doe not correspond to three identifiable organisms) scientists must resort to analogies and metaphors. One of the best is contained in a recent publication, which we will discuss in the section that follows.

Head, Heart or Gut?

It's been said that all science is based on analogies. We use the analogy of a clock to explain the well-organized universe, then continue the analogy by saying it is "wound up" by the Big Banger at the beginning of time. (The book to read is by Princeton's Brian Greene and is titled *The Fabric of the Cosmos.*)

DNA replication is "like a zipper" that opens up and closes in inverse, asymmetrical motions that open it and close it simultaneously.

The cognitive capacity of animals and men are described as having "hardware" and "software" – very analogous to modern computers.

Such is the way with neuroscience – except that the analogies are more like metaphors, because the interaction of the brain with the senses and the emotions that result from the electric signals that convey that interaction is not yet well understood by science.

One very useful analogy is the one used by a trio of scientists (Tara Swart, Kitty Chisholm and Paul Brown) who wrote a refreshingly lucid and recent (2015) book, titled, *The Neuroscience of Leadership.* The authors pinpoint three regions of the human anatomy as competing among themselves for control of our actions.

The three are: "head, heart and gut."

Here's how they explain it:

> *The brain is the master controller of all the systems in the body including the main components of heart and gut too.... So, among many matters we shall touch on, we are certainly going to explore the neurobiology of which system is really in charge: Head, heart or gut?*

That's as succinct as we can get in terms of modern neurobiology. We know that the brain organizes reactions to what the gut feels and what the heart inspires. But which of the three prevails? Which one is really in charge?

Let's continue to analyze these three neurological categories, which are akin to body (guts), mind (head) and spirit (heart). (*Caveat*: when we refer to "heart" in this context, it is not a reference to the actual biological organism. In this sense, "heart" is an intangible substance or component of the human equation – akin to "soul" or "conscience.")

In the process, we will be able to discern important differences between humans and animals.

How Guts, Mind and Heart Interact

Swart, Chisholm and Brown provide remarkably concise descriptions of what they call the "exquisite interaction between nerves and hormones (the endocrine system) in the brain and body as a whole." Here's an example of how the interaction works, when a human feels threatened:

> *The tiniest change in a cell, chemical or electrical signal in the body, gives rise to some compensatory, complementary or inhibitory response in the brain and vice-versa. Like when you feel threatened so your adrenal glands pour out a little adrenalin, then you start to breathe faster and your heart rate goes up. Or when you take a few deep breaths to calm your mind.*

Our species is no longer threatened on a daily basis; therefore, we don't have as much preoccupation with the adrenalin rush that is needed to overcome fear and defeat those who would do us harm. I suppose the exception is sports – particularly the more violent ones like football and boxing.

A more customary preoccupation for the male members of our species is the rush of a substance called testosterone.

The "trouble with testosterone," the trio explain, "starts *in utero* (in the womb) when the embryo with the XY gene gets its first flood of testosterone and starts the journey to becoming male." Yet, the authors continue, it is not limited to men; "women also have testosterone, but much less than men."

How much more testosterone do men have? Glad you asked; men have about eight times as much as women. Furthermore, because metabolic consumption "is greater in men they actually have to produce 20 times as much per day."

The above-mentioned "trouble with testosterone" is discussed in a full chapter later on. We will explore sexual aggression by men (for the most part, including clergy) and the concomitant difficulty in controlling it by every means possible, including societal norms, regulations such as those against workplace sexual harassment, and the various biblical mandates.

The interplay of these so-called passions and legal/ethical constraints is worth a book in itself. It is a major concern of scientific disciplines like criminology, psychology and the ever-present interaction between church and state in a polyglot, democratic society.

Another major concern, which affects both genders almost equally, has reached almost epidemic proportions.

Just as our species has come close to conquering diseases that affect the body, such as malaria and tuberculosis, plus softening the impact of many kinds of cancer (e.g., breast and prostate cancer), the incidence of mental illness cannot be said to improve apace.

In some areas, there is regression, rather than progression.

One mental dysfunction is particularly devastating at this time in the history of humanity. And it is peculiar to our species.

It is referred to as clinical depression.

Dealing with Clinical Depression

Everyone gets depressed – if only for a short period of time. Professor Dingman illustrates that point, when he says that a teenager might say she's depressed because her best friend is on vacation and she "doesn't have anyone to hang out with." Similarly for any "ambitious adult" who might get "depressed that a job opportunity didn't pan out."

Dingman expands on what is our concern here:

> *Depression in the medical sense...is commonly known as*
> *'major depressive disorder," or MDD....People who have*
> *MDD are so melancholic throughout most of the day that*
> *their ability to experience pleasure may be, for the most part,*
> *non-existent.*

People who suffer from MDD "are often plagued by sleep problems (either insomnia or sleeping too much), feelings of worthlessness or irrational guilt, and suicidal thoughts." And the clincher is manifested in the drastic measures they often resort to. Dingman quotes studies which estimate that "depressed patients make up about 60 percent of suicides."

Clearly, it becomes important to understand the causes of depression. Medical science has converged around the theory that the main biological culprit of depression is an enzyme called "monoamine oxidase" (MAO). The first and best known chemical that was discovered to deal with depression was Prozac. Later, other drugs were developed.

"Belle of the Ball"

Dingman calls Prozac the "belle of the ball" and illustrates its popularity, telling his readers that by 1994, it was the second-best selling drug in the world, behind the anti-heartburn medication *Zantac*.

He then proceeds to catalogue the rush of competitive drugs with the same effect, referred to as "selective serotonin reuptake inhibitor," or SSRI. He also documents the concomitant usage in America:

> *Other pharmaceutical companies rushed to develop their*
> *own SSRIs. Soon the market was flooded with drugs like*
> *citalopram (Celexa), sertraline (Zoloft), and paroxetine*
> *(Paxil). Prozac had led the way, but SSRIs in general became*
> *the psychiatric drugs of choice for America in the 1990s and*
> *into the 2000s. By 2005, more than 10 percent of Americans*
> *were taking an anti-depressant.*

In this book we cannot hope to fully explain, let alone "solve" the problem of depression. The science is embryonic and the brain is complicated – to put it mildly.

Our task is to convince the reader that this rather common psychological condition is in many, if not most cases, treatable only with a holistic approach. In other words, I expect that we will ultimately conclude that common clinical depression (as opposed to bipolar depression) is a function of body, mind and spirit.

Assuming that is true, it is also the case that treating depression requires a holistic understanding of our species. And that is precisely what this book is about.

Once again, it begins with accepting the simple, evident fact that humans are composed of body, mind and spirit. This idea is gaining acceptance in many fields of science, particularly those that deal with treating patients who suffer from clinical depression.

"Lost Connections"

One of the most illuminating books on clinical depression was written just a couple of years ago by a person who suffers from the illness, named Johann Hari. He asked himself the key question that has been brewing in my own mind, as I observe a condition that affects so many of my friends and relatives.

Here's his exact words:

> Could something other than brain chemistry have been causing depression and anxiety in me, and in so many people around me? If so – what could it be?

As most doctors will acknowledge, the treatment of any illness, and particularly of any *mental* illness, includes some sort of medication and also includes what Hari calls "a story" about how the treatment will affect the patient.

In other words, you have to convince the patient that a drug will help to heal her, but also that a lot depends on her positive attitude about it.

Hari's insights were formalized into a study that was performed by Irving Kirsch, who became a leader in this field, working out of Harvard University. Together with a graduate student named Guy Saperstein, they performed studies that first divided the patients into three groups: the first group is told they are given an anti-depressant, but it's really a placebo; the second group is told they are given an anti-depressant and actually get one; the third group is not given or told anything – they are simply followed over time.

Comparing the results of their own study with others done in the nation, they found a surprising and very consistent result. Here's the finding, in Hari's own words:

> *The numbers showed that 25 percent of the effects of anti-depressants were due to natural recovery; 50 percent were due to the story you had been told about them and only 25 percent to the actual chemicals.*

The scholarly studies can be summarized into a simple observation: When dealing with depression, the "pep talk" is as important, if not more, than the medication.

And there are myriad other factors at play. Hari lists no less than seven, including the importance, for the clinically depressed, of reconnecting (1) to "other people," (2) to "meaningful work," (3) to "meaningful values" and (4) to "sympathetic joy and overcoming addiction to the self."

I want to take a little time with no. 4. The notion of "sympathetic joy" is quite revealing, as we travel from the purely physical, to the psychological and, ultimately, to the spiritual.

Sympathetic Joy

Hari's book rises to a crescendo of insight in Chapter 20. Here he describes a series of back-and-forth sessions with other subjects, as well as with professionals. The gist of his conclusions can be gleaned from the following passage, in which his friend and colleague-in-depression, Rachel, states:

> *Rachel had come to realize that she was angry and envious a lot of the time. She was embarrassed to say it, because she knew it made her sound bad – but, to give one example, she had a relative who had been driving her crazy for years. She was nice, and Rachel had no reason to dislike her. But her*

*every success – in work, in her family – felt like a put-down
to Rachel, and it made her dislike the relative, and that in
turn made her dislike herself.*

Which is obviously a liberating experience for the person suffering
from depression. I mean, who wants to be jealous of their own relatives –
particularly the nice ones?

And it gets better. Rachel goes on to try the "sympathy" experiment
with people about whom she is neutral, such as total strangers. She describes
rather amusingly how many of us feel about ourselves, in comparison to
others:

> *So you display your life on Instagram and in conversations
> as if you are the "Chief Marketing Officer of Me," not trying
> to get other people to buy anything other than the idea that
> we're awesome and worthy of envy ourselves....*

And the author continues:

> *Rachel didn't want to be this way. Like me, she's a strong
> proponent of skepticism and rationality, so she looked for
> techniques that scientific studies suggested might have some
> basis in fact...And she discovered an ancient technique called
> 'sympathetic joy,' which is part of a range of techniques for
> which there is some striking new scientific evidence...It is, she
> says, quite simple. Sympathetic joy is a method for cultivating
> 'the opposite of jealousy or envy...It's simply feeling happy for
> other people.'*

In the final passage of the exercise, Rachel realizes that there is an even
higher form of sympathetic self-reflection: the one in which you learn to
forgive those who have done harm to you. This sort of meditation "works
particularly well for people who've developed depression as a result of
abusive childhoods," says Hari. And he adds that "they have a 10 percent
higher improvement rate than others."

Loving Your Enemies

Loving your enemies, or being "sympathetic" to them is not an easy exercise. We, who were taught to do that as children, know how difficult that is. And yet we also sense that joy can come from eliminating negative thoughts about others, even if they deserve it.

I recently had to check myself from reacting negatively towards a friend who disagreed with a political decision I had made, in favoring the principal of one project over the friend of my friend, who was an old, frail man. When the man died, my friend blamed me for his friend's death.

That kind of animosity, particularly when it's undeserved, can bring out the worst in us. We can generalize about the offender's ethnicity or his ideology, condemning not only the bad actor but his entire tribe.

A person who suffers from depression that has a clear human cause is particularly prone to blame others; but dwelling on the offender's sins may not actually help heal the wounds caused by that parent, or spouse or former lover.

At the end of the analysis, Hari concludes that the evidence "shows that there are three different kinds of causes of depression: biological, psychological, and social."

Interestingly, the "social" element is close to what most of us consider the "spiritual" component: the idea that we are all worthy of love because we are all children of the same loving creator.

Hari, who is an atheist, acknowledges that there are "other kinds of psychological changes people can try, too. One is prayer – there's evidence that people who pray become less depressed."

Actually, there is tons of evidence that people who pray do better in every respect. But it has to be the right kind of prayer: the kind that is based on love of God and your fellow humans.

Many, if not most mental health professionals accept and implement that kind of holistic approach. As we will see, it represents explicit, scientifically based recognition of humankind as a triune reality.

Holistic Healing

Such is the opinion of Eve A. Wood, M.D., the award-winning author of *There's Always Help; There's Always Hope* and host of a weekly radio show called "Healing Your Body, Mind and Spirit."

The idea that humans are not just a dual-substance (digital) reality, but a three-dimensional (triune) reality, first came to me when I was studying at what is perhaps the best preparatory school in the nation. Not surprisingly, it came from my religion teacher – a very wise and highly educated Benedictine priest named Fr. John Farrelly.

Let me digress and brag a little bit about my high school and the Benedictine monks that taught there. Americans think that the Jesuits of Georgetown and Boston College and so many regional powerhouses named Loyola, Gonzaga and Xavier are the cream of the crop for Catholic academicians.

But if Jesuits have a strong academic tradition, they are newcomers compared to the Benedictines, who preserved civilization and added to it significantly during the early Middle Ages in England and Europe. The ones who taught me, at St. Anselm's in Washington, D.C., inculcated in me a love of both science and humanities, including the study of three languages besides my native Spanish and acquired English: French, Greek and Latin.

That little prep school has achieved national recognition by having the highest proportion of National Merit Scholarships in the nation's capital.

Anyhow, enough bragging. The point is that good old Father John told us one day that our species is really best described as having three dimensions: the physical, the mental/psychological and the spiritual. (Note that the mental/psychological part of us is nowadays generally referred to as the "mind-psyche.")

And that insight has stayed with me, providing compelling explanations for some of the most intractable puzzles of the human condition.

It says here that the triune perspective will ultimately be the key to understanding (and correcting) human deficiencies like depression; criminality; substance addiction; laziness (and its cousin, narcissism); anger and sexual misconduct.

Again, it will take a library and not just one book to carry out that analysis in full. Here I do want to illustrate the triune analysis as it is

53

used by practitioners to treat certain mental illnesses, such as clinical and bipolar depression.

The analysis begins by understanding what we mean by the "spiritual" dimension, which is not necessarily the same as the more traditional concept of "religion."

Let's delve into that. In doing so, my reader will note that it is not a theoretical discussion, but a very practical one. I should add, as an observation that bears more analysis, that academic theoreticians are not as versed in the spiritual dimension as are the professionals charged with actually healing our species.

Score one for the professionals over the professors.

On the Spiritual Dimension of Mankind

Eve Wood defines spirituality as involving "the meaning and purpose of existence." The inspiration may come just as easily from science as from established religions, given that all spiritual traditions "include ideas, values and philosophies about life."

Here's how Dr. Wood describes the interaction of the ideas with the physical brain:

> *When you immerse yourself in those teachings, you process and live them through your thinking, cortical brain. But these faith systems also include actions, such as meditation, song, prayer, yogic breathing, and postures. When you engage in those activities, you involve your deep emotional brain to bring about a response. Involving yourself in spiritual pursuits on a consistent basis is therefore a particularly powerful way to take charge of your emotional life.*

It's important to note that neither the inspiration nor the motivation has to come from any sort of organized religion. For scientists like Deepak Chopra, the intellectual basis for spirituality comes entirely from empirical science – i.e., biology. He argues that the harmonious ways of organic life, in particular, can only be explained through the link that must exist to a divine power – a super-human intellect – that is in unity with all organic life on earth.

Why does Chopra believe that? Parenthetically, he's not the only celebrated natural scientist who does. In other works, I have cited the writings of the person who was perhaps the foremost biologist of the twentieth century: Pierre Paul Grasse. Dr. Grasse edited the 58-volume French Academy of Science's *Traite de Zoologie*, which encompassed just about everything that we know about the animal world. In the process, he concluded that organic life is too complex, too multi-faceted, too dynamically interconnected to conform with the Second Law of Thermodynamics, which says that all matter, left to itself, must disorganize.

Celebrated Columbia physicist, Brian Greene, puts the conundrum this way:

> *Underneath reality's steadfast façade, science has revealed a relentless drama of churning particles in which it is tempting to cast evolution and entropy as embattled characters perpetually fighting for control. The tale envisions that evolution builds structure while entropy destroys.*

Greene goes on to explain that the tale has "some truth to it." Yet, as he explains, there is an "entropic two-step" that allows "structure to flourish here, so long as entropy is expelled there." In other words order can flow from one part of the system to the other as long as the overall amount of disorder continues to grow in a direction towards more entropy (disorder).

That's all very well and good, but by the time the universe evolves to the point that one planet can sustain life, it is no longer an "entropic two-step." *It is more like a dis-entropic cascade* of remarkable order, extraordinarily coordinated, to promote the formation and subsistence of life.

To many scientists, as well as this author, something new is at play in organic life; and physics (the Second Law of Thermodynamics) tells us it must be a force from outside the system. This outside force takes a whole new dimension as we move from animals to humans; later on we will discuss that particular phenomenon, which begs for a metaphysical explanation.

But before we do that, let's delve a little bit into the pre-human, organic world, where order and complexity seem to grow, rather than decrease in time. The phenomenon has a complicated, multi-syllabic name. Hopefully, I can make it easier to understand for my reader than it was for me, when I first read about it.

It is called "synchronicity."

Synchronicity

Like Chopra, Grasse concludes that there is an organizing force in nature that begs for a higher power at work. By itself, nature has an irrevocable tendency to dissemble and become less organized. Gas escaping from a Coca-Cola bottle does not ever come back in; eggs that fall off a table and break don't ever reassemble themselves and form an egg from the spilled components.

Yet most living organisms, if not dismembered or killed, function like well-oiled machines. The precision that it takes for the many cells of an organism to act in unison (synchronously) is extraordinary and extraordinarily difficult for empirical science to figure out.

This trait, which is the key distinction between inorganic and organic life, is referred to as "synchronicity." In animals, it is a phenomenon that allows flocks of birds to react simultaneously to a sudden wind or other danger. The coordinated reaction is so fast that it is impossible for even an electromagnetic signal (such as light) to reach all the other birds.

In that case, scientists are forced to conclude that the animal hardware is so well crafted that it responds with the identical reaction, to the same stimulus, at precisely the same time.

As difficult as synchronicity is to explain when a flock of birds reacts to a well-known atmospheric phenomenon, it is much harder to explain how human beings act in unison in response to a previously unknown set of conditions.

How can eleven football players, or five basketball players, coordinate the movement of the ball so synchronously, moving in a split second in such a harmonious way? How can a million commuters choose the fastest and safest way to get to work, when considering so many alternative routes? How can society function when millions of people are choosing to produce and consume without coordinating their invidual decisions?

As always, science pursues even the most difficult questions with confidence that a cause-effect relationship can be discovered. It may not be something that we can expect to ever replicate in the laboratory, but at least we can observe the phenomenon at work, using instruments that allow the tracking of neurological signals, as they leave and return to the brain – often with instructions, transmitted by the nerves, for some glandular or muscular motion.

In recent years, thanks to modern technology, we have a new term for this kind of research. It is called "applied neuroscience."

Applied Neuroscience

As the previously mentioned trio (Swart, Chisholm and Brown) explain, applied neuroscience "is a relatively new field" that "aims to bridge" the gap between what the brain does and what the human, whose brain it is, does. And, they add: "The elements of all brains are the same but the way they have been organized within any one particular person is unique to that person."

Clearly, the human mind-psyche is unique. It is so unique and so complex that we can never replicate it. Every effort to build a robot that can take the place of a human person soon reaches a frustrating dead end.

You can copy a person and assemble all the parts together – you can make a robot that has all the components of a human, but you can't make it act like a human.

Why is that? How do we explain that for human beings, the whole is so much greater than the sum of its parts? Put another way: Why can't we humans build a robot that caresses like a mother, or holds the hand of a child, or smiles to convey the kind of joy that will lift the spirit of another member of our species who is momentarily joyless?

Let's start with the chemical analysis and see how far we get.

The Cuddling Hormone

The human brain is part of the explanation. I did not know this when I started writing this book, but it turns out that there is a specific hormone that is programmed to provide all the amazing effects of cuddling.

The hormone is called "oxytocin" and recent research associates it, in the case of women, with contractions of the uterus needed for birthing. For both sexes, it appears to enhance the instinct to cuddle with one another and with infants.

As happens with astrophysics and atomic physics, there are obvious limitations to the reach of science when it delves into these kinds of human traits. Something in our species speaks way beyond the assembly of parts that adds up to a whole breathing, smiling, loving person.

Again, we can supplement missing hormones, and repair the nerves, glands and other sensory channels that interact in the human person, but we know, instinctively as well as analytically, that we cannot replicate a human being by assembling the parts exactly as we see them in well-functioning humans.

And what exactly is a well-functioning human?

Well, the simple answer is that to be fully human is to have a high level of what has traditionally been called "free will."

Neuroscience and Free Will

The study of the human brain and its interaction with nerves, glands and outside stimuli (physical and psychological) takes place in both the empirical lab (through measurements of electric impulses) and in the sociological lab (through statistical studies of human behavior).

Unlike other areas of science, such as anthropology, scientists in this field don't doubt that humans exercise a measure of free will. Here's a passage from the aforementioned *Neuroscience for Leadership*:

> *That human will or will power (defined here as the determinatoion to pursue a chosen action or goal) exists has been demonstrated by experiments pioneered by Roy Baumeister....A key manifestation of will power is self-control or self-management, the ability to resist instinctive or impulsive desires, to temper or reframe intuitive reactions, regulate emotions and feelings, remain focused on an activity, thought or goal, however long term, and balance*

both emotional and rational/logical (so called hot and cold
systems) processes in decision making.

Neuroscience is merely affirming what the best and brightest in history have proclaimed. Throughout history, our best thinkers have argued that humans have a measure of free will that controls our negative impulses, rationalizes them and enables the brain to delay gratification.

It wasn't just prophets and priests and rabbis, it was also the great philosophers. Nowadays, it includes psychologists, neuroanatomists, criminologists and jurists. Pretty much everyone except sociobiologists.

Drawing partly from logic, partly from introspection (and perhaps with a touch of inspiration from what were thought to be divinely inspired writings), the great thinkers shaped our notions of what ultimately came to be called the "natural law." (The historical and philosophical basis of the natural law is explored more fully in Chapter VI.)

Throughout history, the greatest minds have concluded that human personhood includes an outside code of conduct – some notion of what the natural law requires of us that is beyond our appetites. There is some intangible element, some sort of out-of-body phenomenon at work that reflects a tension between our will and our instincts.

For almost all classical thinkers, the logical assumption was that free will was real and that the natural law was also real. The question, then, was: How do humans adjust their freely made decisions to some sort of moral code? And the related question was: Where do humans find that moral code?

For most of recorded history, that sort of talk was left for philosophers and theologians, in part because it was assumed that the natural law was imposed by a supernatural being. And it was also assumed that the supernatural being had spoken clearly through the Ten Commandments.

Nowadays, the study and treatment of anti-social behavior (i.e., behavior contrary to what our nature commands) is very much a part of empirical science – particularly the healing sciences. For psychologists and psychiatrists, for addiction therapists, for marriage counselors, for any professional dealing with teen-age runaways or those attempting suicide, the idea that there is a moral barometer lurking inside our beings, and possibly inserted there by an infinitely wise creator, is a tool of the trade.

The transcendent being, the all-wise creator, doesn't have to have a beard or dress like the God of Abraham. More commonly, in most civilizations, it is a unifying principle – a powerful, all-loving parental figure who never gives up on his/her creatures, always forgives past mistakes, always is ready to embrace the person who wants to return to his/her family and receive their warm embrace.

It's all about the spiritual realm; and it doesn't contradict science.

Nowadays, it seems quite compatible with science – particularly the therapeutic science we call psychology.

From Freudian Psychology to Numinosity

At the beginning, psychology was seen as a totally secular science in which the idea of God was, at best, a nice, self-acquired trait – a crutch to carry our species through difficult times of material or emotional deprivation.

More recently, psychologists, as well as socio-psychologists who study group behavior, are discovering what is for most of us an obvious nexus: that which exists between faith in a creator and psychic health. There is even an archive of scholarly studies and articles dedicated to the topic.

It is called the "Archive for the Psychology of Religion." And it contains a multiplicity of well researched papers. One such "Research Report" was just published in 2019; it was entitled "Evaluating the relationships among religion, social virtues, and meaning in life." Three co-authors are listed: Neal Krause from the University of Michigan, Peter C. Hill from Biola University and Gail Ironson from the nearby University of Miami.

Having examined an extensive bibliography, the authors readily conclude that "a number of different facets of religious life have been linked with meaning making." That conclusion is followed by an impressive list of examples:

> For example, some investigators propose that a sense of meaning arises from participation in religious rituals, including worship services. In contrast, other researchers maintain that meaning is found in the process of developing a deep sense of commitment to religion, while yet other scholars

argue that meaning emerges by adopting security-focused religious beliefs (e.g., the belief that God watches over the faithful). In addition to this research, some investigators report that positive emotions are associated with a deeper sense of meaning in religious settings, while other researchers demonstrate that people who are more likely to take the perspective of others and simulate their thoughts, feelings and intentions (i.e."mentalize") are more likely to derive a greater sense of meaning in religious contexts. [Citations omitted.]

The authors then proceed to identify "beliefs that are shared by every major faith tradition in the world." They categorize them as three, key "social virtues" which are roughly defined as follows: compassion, forgiveness of others and a willingness to provide "social support for others."

Two of the mentioned authors' "social virtues" correspond almost exactly to my three faith-base traits. "Compassion" is equivalent to "faith-based altruism." "Forgiveness of others" is what I call "faith-based tolerance."

"Social support for others" is roughly equivalent to the key leadership trait that I refer to as "faith-based persistence." My term sounds a bit more ego-centered; however, because it analyzes how leaders are resolute in overcoming seemingly unsurpassable obstacles in the battle for communal goals, it can be seen as a trait with particular importance to the wellbeing of society as a whole.

Perhaps for that reason, the trait is needed, and displayed best, by the alpha member of the species – the leader.

Yet it is key to any human engagement, whether it be sports or work, which requires overcoming failure, fatigue, and fear. In other words, pretty much every human activity that is intensively productive – that is inclined towards excelling.

Faith-based persistence is possessed by all humans to some degree, but is essential to a leader.

But before we delve into the high-performance side of the human mental-psychic spectrum, let's consider the opposite side of the spectrum, i.e., the dysfunctional side. For obvious reasons, there is a lot more scientific data describing psychological dysfunctions than psychological normalcy. The healing sciences are are all about curing people....

In that pursuit, modern science has begun to accept the possibility that many psychic dysfunctions can be, if not treated, at least alleviated by faith in God. As previously noted, this notion has gained acceptance particularly among practitioners of the healing sciences; it is not so prevalent among theoreticians.

In therapeutic mental health, a strong correlation has been found between psychic health and the yearning that most in our species have for a more perfect form of happiness – the complete fulfillment of which may have to wait for an afterlife. Psychologists and other mental health professionals are increasingly conscious that deficits in contentment, whether related to addiction or genetically depressive personalities, can be alleviated by reconnecting with some sort of infinitely loving being.

Let's explore those advances in mental-health therapy.

The Spiritual Component of Mental Health

The interface between the physical and the spiritual dimensions in humans is particularly compelling when dealing with those who are suffering some sort of psychological ailment. Various practitioners of psychiatry and psychology have endeavored to explain how crucial the holistic approach to health care is.

Here's a passage from one of the most celebrated, Thomas Moore, who wrote a popular best seller entitled *Care of the Soul in Medicine.*

> *You see a patient lying in a hospital bed and think that his illness is all there on the bed. But if you could crawl into his head, you would find concerns about the family, the house, the neighborhood, work, the car, perhaps even international politics.*

Moore is definitely onto something. He is not some kind of modern faith-healer. He acknowledges that he is alive "because of the wondrous technology of angioplasties and stents" and that he has "submitted to standard medical expertise for tonsillitis, appendectomy and angioplasty."

He readily acknowledges that he takes "several pills each day for high blood pressure, cholesterol and thyroid."

Yet the physical care of the body is incomplete by itself. Moore argues that:

> *"The soul of medicine...entails...effective dealing with emotional and personality issues...a multidimensional view of what a person is, and the capacity to address spiritual issues of meaning, anxiety, vision, hope and comfort."*

It is evident from the writings of these health care professionals that modern science is now talking in the same basic terms as the philosophers and theologians of old. When they say "multidimensional" what they mean is another dimension beyond body and mind. And when they refer to "spiritual" values, they are talking about religion, although they don't particularly like that word – probably because it might make the reader think that they are referring to *organized religion*.

It's somewhat ironic that scientists don't like to mention "religion," since the word comes from the Latin for "reconnect." And modern psychology is all about "reconnecting."

Let's consider the insights of the previously mentioned Dr. Eve A. Wood.

Science Tackles Spirituality

Here's how Dr. Wood explains the relationship between spirituality and reconnection:

> *Spirituality encompasses the experience of connection and oneness that unites all creatures — that healing force that reverberates between two or more people, or between a single soul and the universe. This energy is also sometimes apparent in an astonishing moment when we feel that the wisdom of the world has been revealed to us. Spirituality is the whole, which is infinitely greater than the sum of its parts. It's the holy or the miraculous in the mundane.*

There's much to digest in that single quote. The balance of this book is dedicated to that idea. In the coming years, scientists will be studying how the spiritual dimension helps in the treatment of mental illnesses like clinical depression. Some call it "holistic medicine." Others refer to the integral approach as "wellness" rather than just "health."

In effect, the health sciences, particularly the therapeutic ones, are embracing the three-dimensional reality that is part of human nature. Increasingly, the professionals who seek to heal the ailing mind-psyche of their patients, acknowledge that humans are enormously dependent on healthy interaction with others of our species.

In summary, material wellbeing can only get you so far, in the case of humankind. *You can feed a child; you can educate a child; you can clothe and house a child. But if you don't love a child, you will end up with an atrophied being.*

The psychic component in physical health has been accepted for decades now. The phenomenon is called "psychosomatic," which combines "psyche" with "soma," which is the Latin term for the body. But now we are entering a new horizon, where psyche and body meet phenomena from outside the individual.

It is increasingly clear that humans draw nourishment from other humans – and from another, transcendent being whose infinite love and wisdom beckons the human soul like a Pied Piper beckons his followers.

Humans can never be alone for too long. They hunger for the company of others of their kind and, at times, for a supreme being who occupies a totally different dimension that cannot be touched physically or measured with a yardstick, but that is vital to our happiness on this earth.

The "relational" dimension in humans is the thing that most distinguishes our species from the animal world. Undoubtedly, animals communicate with each other about dangers or pleasures around the corner. They cuddle and tickle each other and clean their furs. They lament the death of a member of the family or tribe.

But animals are basically self-centered. That was one thing that Darwin got right: Animals seek to do two things: survive and procreate.

Humans teach one another; convince one another; inspire one another.

We humans are, to a great extent, shaped by our relationships as much, if not more, than by our bodies and our upbringing.

Based on the above, and on the combined wisdom of the best minds throughout recorded history, it is hard to conclude that our species evolved spontaneously the oh-so-significant traits that makes us social beings. To me it is self-evident that our species at some point obtained, from an external source, the ability not only to think, but to communicate at a level that is way beyond instinct to survive or procreate.

To socialize.

When psychologists refer to such social traits as "relational," what they mean is that humans are particularly prone to seek happiness by interacting with each other. At some point in our lives, the material comfort we seek, even when our own cup "overflows" is not sufficient to fill our beings. We need to fill our psychological cup with a whole different kind of sustenance.

Humans who are reasonably satisfied as to their own personal needs inevitably try to find meaning in helping others, even if they are total strangers. Scientists refer to that as "altruism." Most of us refer to it as "love."

The sociobiologists do their best to fit this trait into their Darwinian paradigm, as if it were yet another example of how humans evolved in order to survive. In reality, it is a trait that points in a totally different direction: towards what Elton John referred to as "affairs of the sky."

Altruism = In the Image and Likeness of God

This trait, called altruism by scientists and love by average folks, has little or nothing to do with survival of the species. To the contrary, it is a trait that appears to borrow from the maker of all other beings the desire to make other beings happy.

As we shall see, it also compels a triune understanding of the Creator.

God, if he/she exists, is by definition self-sufficient. But God is also an all-loving being. Otherwise why would God create beings?

Now, let's take those two premises – that God is self-suffient and also infinitely loving. Doesn't that sound like a contradiction – an oxymoron?

No matter how hard you try to slice it, there is only one way to harmonize those two traits. The logic of the solution was best expressed by theologian H.P. Owen, who said: "The doctrine of the Trinity reconciles the paradoxical affirmations that God is self-sufficient and that he is love."

I am not asking my reader to accept that the infinite, all loving Big Banger is somehow a three-person entity. I'm only asking that my reader hold that thought, as a mere possibility, and see if the logic of it helps analyze our species.

Certainly, from the purely scientific standpoint, there is plenty of evidence that humans, particularly those with severe addiction problems, need to believe that there is an infinite being out there that loves us, despire our addiction.

And addiction is not the only therapeutic health condition that is relieved, if not cured, by a strong dose of spirituality. We have already talked about clinical depression. Later we will reach the same conclusion as to narcissism and its close cousin, sociopathy.

In almost every aspect of the most modern and holistic analysis of human function and dysfunction, we are confronted with three dimensions – the physical, the psychological and the spiritual.

Yet that rather evident view of our species is not shared by many in academia. It is particularly so among anthropologists and their materialistic cohorts in academia.

The idea that humans respond to stimuli that can be described as body, mind and spirit is simply rejected by anthropologists of the school referred to as "deterministic" or "reductionist." These folks are convinced that there is only one dimension in reality – the material.

For that reason their philosophy is described as scientific materialism and they are aligned, in terms of anthropology, with the previously mentioned sociobiologists.

Let's review the essential tenets of scientific materialism/sociobiology and compare its logic and its tenets to the integral humanism that is the basis of this book.

Keep in mind the simple math: Scientific materialism stands for one dimension in reality; integral humanism argues for three.

An Opposing View: Determinism and Reductionism

There is a strain of biology and anthropology that are like the strict constructionists in the law, in that they follow very strictly the two Darwinian tenets - survival and reproduction. Because they don't believe that humans have any choice in how they behave, they are referred to as biological "determinists."

Philosophers describe biological determinism as "reductionist" because it sees all reality as defined by, or "reduced to," one dimension, which is the material. In the case of humans, all motivaton is reduced to the instinct to survive and reproduce.

The human condition is so much more complex, so much more intriguing and exciting that it makes you wonder what art, what movies, what song the determinists like.

As far as movies, I know what most of us like: "Casablanca." We love to watch Humprey Bogart and Ingrid Bergman struggle with romantic love while also struggling with the demands of country and liberty.

They are complete human beings – much different from what the sterile anthropologists, or behavioral determinists (aka, sociobiologists), portray.

Clearly these folks don't put much stock in romantic love. But what about sex? My reader might be amused to see how sociobiologists stumble through the analysis of what is perhaps the most talked about of all human traits.

CHAPTER V
SEX, GENDER, SCIENCE

The 'sex essentialist' view interprets sex as a fundamental category that divides humans and other sexually reproducing organisms...neatly into two types. Bennet McIntosh, *Harvard Magazine*

To the sociobiologists in academia, the fact that humans have a sexual dimension is either a necessary fact of procreation or a puzzling complication that provides a transition from asexual organisms to mating, coupling, voraciously aggressive animals that ultimately evolved into us.

To the criminologists and the marriage counselors, the sex drive is often the cause of mayhem, improper conduct, and horrendous abuse of minors.

To the artists, composers, screenwriters and novelists, it is the stuff of life. The same is true for many if not most marketing gurus, as they sell us soap, beer, cars and cosmetics.

To the priests, ministers and rabbis, sex is a complication – a mostly forbidden fruit whose prohibition merits not one, but two of the ten basic commandments of a moral life.

The excessive, unbridled instinct to copulate, particularly but not exclusively by males of our species, is a cause of great concern for the mainstream religions. Over the course of history, each important religion has taken a strict view of the proper use of our sexual powers.

Judaism, Christianity and Islam have all couched discussion of sex in normative terms. *Thou shalt not commit* adultery. *Thou shall not covet* thy neighbor's wife. *Thou shall cover* thy nakedness.

You really can't blame the religious leaders. For the entire history of organized religion, they have had to combat those who argue that sex is an unavoidable force and should not be repressed.

The high point of that school of thought came in the middle of the last century with the research and writings of Alfred Kinsey.

Sexual Liberation a la Kinsey

Alfred Kinsey (1894-1956) did his best to convince Americans that cheating on spouses was really the norm – with half of all men and forty percent of women being unfaithful.

As Ben Shapiro describes it, the new sexual morality prescribed that "human beings could be bettered by casting aside the vestiges of the old morality." Continues Shapiro:

> *And the best news of all was this. It was all natural. No more struggling to seek the natural law; no more utilizing reason to hem in biological urges.*

I remember that during the sixties, when even the more prudish among us had almost (not quite) accepted the idea that science had given its blessing to free love, the popular saying became: "If it feels good, do it."

Shapiro connects that saying to Jean-Jacques Rousseau's concept of the "noble savage" who only needs to follow his/her instincts, devoid of any moralistic constraints. It was a great and convenient rationalization of the Kinseyan free-love idea; here's Shapiro:

> *If it felt good, not only should we do it, we had a biological imperative to do it. Forget striving for existential meaning – we would all find truth by freeing ourselves.*

The era of sexual liberation put religious leaders on the defensive. They were convinced, from biblical sources, that the ideal form of sexual expression is in a properly constituted marriage. Furthermore, they sensed, even without the laborious studies we have now, that monogamy gave stability to the entire family unit.

That theme will be repeated in other passages of this book. I will argue that sex between adults of our species, with the intention and effect of procreating and raising offspring, is best experienced within a stable family unit.

This idea is quite a departure from the sociobiologist's thesis that the sum total of human welfare – the be-all and end-all of our existence – is to survive long enough to procreate. As with other aspects of modern materialist thought, that interpretation of the reason and the value of sexual attraction is as sterile as trying to explain a painting based on lines and angles alone.

It is woefully one-dimensional. It shortchanges our species. And it misses the point entirely, as we shall see.

But let's take a closer look: My reader might be surprised at just how sterile, if not downright shallow and unconvincing, their theories are.

Sex: A Dead End to the Evolutionary Biologist

Evolutionary biologists have yet to find a satisfactory, Darwinian explanation for sex. *They can't even explain why there are two genders.* For them, the only instinct that counts is the one to survive and procreate the species; and that, in theory, would be easier and more straightforward if there was only one gender to reproduce.

In effect evolutionary biology is currently unable to give a reason why sex matters.

Tell that to my readers!

In succeeding chapters, I will treat the concept of human love and human altruism. I will argue that it is more than a concept; it is a vital reality that distinguishes our species from all others that preceded it.

Enveloped within that reality, which prompts mature members of our species to subsume their instincts to a higher purpose, is a powerful motivator, which has nothing to do with survival or procreation.

Romantic love has that intangible, physically immeasurable quality.

But let's not get ahead of ourselves. Let's see what the empirical scientists say. In that vein, I began this chapter with a title and a narrative derived from what is the best known university in the world – Harvard.

The article is by Bennett McIntosh, but its subject is the author of a forthcoming book titled *The Maternal Imprint* (University of Chicago 2020). The article has a rather cryptic title: "Sex, Gender, Science." It's almost as if the author doesn't want to even hint at what is in the text.

As it turns out, there is not much.

The Science of Sex, According to Harvard

The cover story was carried in the November-December (2020) edition of *Harvard Magazine*. Its topic was the writings of Harvard's historian and philosopher, Sarah Richardson. Interestingly, the index to the prestigious magazine referred to it as being related to "the science of sex."

And so we shall do our best to deal with it – at least initially. And that begins with basic biology. Professor Richardson explains the phenomenon in its most basic, molecular terms:

> *The story told in introductory biology textbooks is relatively simple: each set of parents confers 23 sets of chromosones on each child – 22 of which are matched pairs and two of which, the X and the much shorter Y, determine sex. Males have an X and a Y, while females have two Xs, and from this all the other hallmarks of sex – gonads, hormones and genitals – follow.*

So far, so good: Science and religion are on the same book and page.

However, the above classifications are not universal. There are men with an extra Y chromosome, and they tend to be taller – which perhaps explains why they have a higher chance of engaging in criminal conduct. (For years, they were referred to as the "supermales," until it was discovered that men with an extra X chromosome also tended to be taller and commit more crimes.)

As can be seen from the above, focusing merely on biological traits is not only frustrating, but boring. The antiseptic treatment of sex by the sociobiologists, as we shall see, starts off boring and only gets worse.

On the Sociobiology of Sex

Human sexuality, for many anthropologists, is not an ennobling trait. Nor is it in the least romantic.

The quandary for evolutionary biologists is that sex is not necessary for evolution; worse than that, it is an impediment! So why did animals evolve from asexual, replicating entities to wonderfully sexual beings?

This question poses a quandary for evolutionary biologists; and the quandary extends all the way to our species.

In another book by this author, I catalogued the various theories advanced by evolutionary biologists to explain why nature even evolved two different genders in primates, including human beings. The analysis bears reading; it shows a veritable cacophony of strange correlations that lead, in every case, to a dead end. The existence of a male sex does not seem, in these recent scientific studies, to confer any evolutionary benefit on the various species that have more than one gender.

The typical evolutionary biologist, working under the assumption that humans are just like animals, looks for ways in which the instinct to survive maximizes life span and, consequentially, the number of sexual couplings which support natural selection. They soon run into all kinds of logical and logistical walls.

Here's a sample from Jared Diamond's *The Third Chimpanzee:*

> *For sexual selection to work, evolution must produce two changes simultaneously: one sex must evolve some trait and the other sex must evolve in lockstep a liking for that trait. Female baboons could hardly afford to flash red buttocks if the sight revolted male baboons to the point of later becoming impotent. As long as the female has it and the male likes it, sexual selection could lead to any arbitrary trait, just as long as it doesn't impair survival too much.*

All of this thinking totally breaks down when discussing how humans come to mate. According to Diamond, all rigorous studies show that "people tend to marry individuals who resemble themselves in every conceivable character, including hair and eye and skin color." He continues:

*Thus, if you're fair-skinned, blue-eyed blond who grew up
in a family of fair-skinned blue eyed blonds, that's the sort
of person whom you'll consider most beautiful and will seek
as a mate.*

Well, yes, those studies are probably accurate; I mean, people tend to
marry those closest to them geographically, and we've already admitted
that those tend to have similarities in hair, eye and skin color – though
those traits don't seem to provide any greater chance of survival.

Diamond almost admits that the search for a cause-and-effect relation
between sexual differences in humans and survival of the fittest has not
been scientifically successful; he says, for example:

*A visitor from Outer Space who had yet to see humans could
have no way of predicting that men rather than women
would have beards, that the beards would be on the face
rather than above the navel, and that women would not have
red and blue buttocks.*

You're not going to hear me even touch that last observation. I have no
idea of the colors of women's buttocks; and if I did, I could not comment
on the colors of any woman's buttocks, other than my wife's – who,
parenthetically, would not want me commenting on hers even in an
academic exercise like this.

When you read these kinds of comments by renowned scientists, the
song that comes to mind is one that says: "What's Love Got to Do with It?"

Well, love, in the case of humans, has a lot to do with choosing a wife
or husband. If not, love songs would not sell so much.

On the Splendid Reality of Romantic Love

Sometimes it's necessary for us laymen and laywomen to stand up
for the things that matter and ignore the strange ideas of some scientists
who have maybe spent too much time languishing in their lugubrious
laboratories.

For me, as for the average, run-of-the-mill guy out there, sex –
particularly when combined with love – is a "many-splendored thing." It

has many dimensions. It is as complex and contradictory a phenomenon as altruism.

It makes grown men cry and brilliant women crawl.

Sex is the stuff of movies, plays, books and at least a third of all the conversations that both genders have. (It is estimated, in the case of men, that we think about sex an average of fourteen times a day!)

There is no way to overstate the importance of romantic love, and its sexual expression, to both the individual and the society of individuals. Of all human relations, the bond between lovers, particularly in the traditional family unit of husband, wife and offspring, is of paramount importance. This is not a religious observation; it is not a biblical affirmation; it is not a romantic illusion.

It is a self-evident fact.

The entire fabric of Judeo-Christian civilization is grounded on the strength of the bonds that bind the family. And the strength of the family is grounded on the strength of the bonds that bind the parents. I will devote a chapter to that sociological analysis. (See Chapter VII.)

Any psychologist, any sociologist, and any educator can bear witness to that.

In terms of public policy, it is also quite evident that a stable family, preferably composed of father, mother and offspring, is the first block in the societal edifice. And sexual expression is vital to a marriage. The world's literature recognizes that phenomenon, as it hopefully grows from the day the couple feels attracted to one another.

This simple truth evaded some theologians, who historically have seemed obsessed with lessening the importance of sex and, in particular, with prohibiting sex outside of marriage. More recently, however, even the most orthodox of Catholic popes have begun to exalt the importance of sex.

This is good news for observant Catholics. It is particularly good news for the female gender, many of whom were traditionally shielded from the very idea that sex could be pleasurable. Could it be the case that the more religious women are turning the tables on the more licentious ones, in the area of sex?

The "Revenge of the Church Ladies"

A study performed by the University of Chicago at the end of the last century confirmed earlier studies at Stanford and elsewhere, which suggested that religious couples have not only a stronger marriage bond, but even better sex.

I'm sure this finding surprises my reader. But it should not. Let's review some of the reasons gleaned from the study by the reporter (William R. Mattox, Jr.) who chronicled its findings for *USA Today* (February 11, 1999).

Here's an excerpt from the article:

> *[T]he Chicago study was hardly the first to show a linkage between spirituality and sexuality. In fact, a 1940s Stanford University study, a 1970s Redbook Magazine survey of 100,000 women and at least one other study from the early 1990s all found higher levels of sexual satisfaction among women who attend religious services.*

A variety of factors are thought to contribute to this correlation. The author cites a 1993 Janus Report on Sexual Behavior which found that the nonreligious "have a tendency to focus on the more technical or physical performance aspects of sex, while the religious pay more attention to the mystical and symbolic dimensions of one's sexuality."

And, adds the author, "churchgoers are apt to delight in the Edenesque pleasure of 'being naked and not ashamed' of celebrating the 'transcendent intimacy' found only in the marriage bed."

It is significant that the studies point, in particular, to the difference between the sexes. As observed by sex therapist Mary Ann Mayo, most major studies show that the connection between monogamous marriage and sexual satisfaction is particularly strong for women, since "their sexual responsiveness is greatly affected by the relational context in which lovemaking takes place."

That is hardly surprising. The integral or holistic understanding of the human reality often appears more of a feminine than a masculine trait. Some thinkers argue that such a holistic grasp of reality extends beyond sex and marriage to theological truths. (The book to read is Karl Stern's *Flight from Woman*.)

Catholic theology has also evolved radically towards an affirmation of sex that would have shocked Augustine and Aquinas. Until recently, the Church has referred to sexual intercourse as something primarily important to reproduction of the species. (In that sense, ironically, mimicking the Darwinian concept of natural selection....)

But the church has become more enlightened. Sexual attraction as part of a well founded marriage, is now considered an essential component of family life.

As we shall see, Pope Benedict XVI waxed poetic on the topic.

Eros As Divine

Here's how Benedict XVI in his encyclical. called *Deus Caritas Est* ("God Is Love"), describes the progression, which in the old days treated sex as an instinct to be controlled and now recognizes sex as a sharing in some way with a divine form of ecstasy:

> *That love between man and woman which is neither planned nor willed, but somehow imposes itself upon human beings, was called 'eros' by the ancient Greeks.*

Pope Benedict proceeds to trace a history in which the Judeo-Christrian tradition seemed to initially treat *eros* (sexual love) as a sort of *necessary evil* and slowly evolved to transform it into a *necessary good*, conducive to both procreation and strong marriage bonds. Says Benedict:

> *According to Friedrich Nietzsche, Christianity had poisoned eros, which for its part, while not completely succumbing, gradually degenerated into vice. Here the German philosopher was expressing a widely-held perception: doesn't the church, with all her commandments and prohibitions, turn to bitterness the most precious thing in life? Doesn't she blow the whistle just when the joy which the Creator's gift offers us a happiness which is itself a certain foretaste of the Divine?*

Clearly, Benedict is acknowledging at least a widespread perception that Christian teaching casts sex in a negative context; but he proceeds to counter that and simultaneously offer us a foretaste of a refreshing exaltation that has characterized the last half century of both Church doctrine and pastoral teaching.

In his words:

> *Did Christianity really destroy 'eros'? Let us take a look at the pre-Christian world. The Greeks – not unlike other cultures – considered eros principally as a kind of intoxication, the overpowering of reason by a 'divine madness' which tears man away from his finite existence and enables him in the very process of being overwhelmed by divine power, to experience supreme happiness. All other powers in heaven and on earth thus appear secondary.*

Benedict acknowledges that the Old Testament rejected what he calls the "extreme perversion of religiosity" that "found expression in fertility cults, part of which was the 'sacred' prostitution which flourished in many temples."

Fair enough. The Jewish tradition did a lot to liberate women from cults that treated them as fertility objects and the more widespread, societal belief that they were inferior to men. Moses's Decalogue emphatically commanded respect for wives and, by derivation, for women generally. And the Christian tradition enhanced that by promoting monogamy and faithfulness "until death do us part."

Furthermore, both traditions emphasize the non-physical dimension of our species. In other words, the existence of a "soul." Note, however, that for many centuries, there has been an undercurrent of tension between the body (flesh) and the soul – at least where it concerns sex.

Body v. Soul in Antiquity

Benedict tells of a legendary encounter between Gassendi, the epicure and Descartes in which the former engaged in the humorous greeting: "O Soul!" to which Descartes would reply: "O Flesh!"

Although Benedict would not admit it, the Church has over-emphasized the primacy of the soul over the flesh; frankly, it oozes out of the writings of St. Paul, to which the Church devotes one out of every three biblical readings each Sunday during mass.

For us who grew up as Catholics, there are many examples of this exaggeration, which is actually – in its extreme manifestation – a heresy known as Manicheaism. Under that philosophy, humanity is seen as a "byproduct" of a battle between man as God's proxy, and the devil.

That is not the Judeo-Christian view at all. Man, in all his physical traits, is good from the start and should be good to the end. But it is ineluctably true that the human sex drive is a maddening force which often wrecks not only marriages, but the emotional lives of a huge proportion of our children.

The nuns that taught many of our Catholic girls would often go beyond the strictest condemnations of illicit sex by St. Paul to absurdly comical prohibitions. One of those that became famous during my years in Washington, D.C. was that girls, according to some "Mother Superior" in some Catholic high school somewhere, could not be permitted to wear shoes so shiny that they would reflect the light and show the reflection of their underwear to the boys!

Those days are gone, and Pope Benedict, as well as his successors, all emphasize the idea that romantic love, far from being a forbidden intoxication that prompts men to use women as sex objects, is a "single reality" that seeks to give as much as to receive and that treats the beloved as an infinitely worthy person.

What's more, current church teaching suggests that romantic love is about the closest thing that humans can experience to the love that exists within God himself/herself.

Failing that kind of understanding and that kind of mature handling of what is admittedly an intoxicating appetite, the human sex drive can be as destructive as any other human trait.

That is made evident by the current state of child abuse.

The Sex-Abuse Scandal

In America, the media concentrates on incidents of sex abuse by priests, actors, and other wealthy and powerful men. Those incidents are scandalous, of course, and worth rooting out with harsh punishment meted out to the transgressors.

But it is merely the tip of the iceberg.

Recently I was visited by the executives of a Miami social agency called "Christy House." This institution is named after a nine-year-old girl who was sexually abused by her father.

That scenario is more frequent than one would imagine, but it is not nearly as frequent as sexual abuse by step-fathers, uncles, boyfriends and others in the extended-family household where many of our children grow up.

The statistics adduced by the professionals in the field are shocking: One out of four girls and one out of five boys are sexually abused at some point in their lives!

Our task here is not to find prescriptions for the problem as much as to understand it. What is it in our species that provokes the abuse of children by adults who are otherwise, in the main, law-abiding? Why are so many men, and a somewhat smaller number of women, so inclined to trespass beyond their own religious views, and what is generally understood to be the natural law?

It all starts by recognizing that humans do, indeed, have a soul, or a moral barometer of some sort. For Jews and Christians, the corollary to that is the existence of another, less disciplined component.

The overwhelming majority of thinkers and scientists are convinced that our good angels are in control most of the time. And there also seems to be a consensus that humans are guided by a "conscience" (Freud's "super ego") that impels us to be kind, tolerant and respectful toward others.

Where there is not much of a consensus is on what propels humans in the opposite direction. The instinct to survive is clearly a motivator, when we are physically attacked or deprived of the basic means of survival, such as food or water.

Next in line, as far as traits that cause a certain amount of aggressively anti-social behavior, is the instinct to mate. Psychology has advanced a great deal since Freud tried to convince us that all men dream of having sex with their mothers. That is clearly a pathological interpretation that may say more about Freud than about the rest of my gender.

Even so, there is no doubt that many, if not most, men think about sex often. And there is no doubt that failure to control the human sex drive leads to all kinds of pathological behavior. We have already discussed the sex-abuse scandal and the damaging effect of broken families.

A more in-depth analysis, by people more qualified than I, needs to be done on this topic; and it needs to be done in a way that recognizes the three dimensions of our species: body, mind and spirit. Sex within the context of romantic love for a spouse, with the possibility of begetting children, is too important a phenomenon to be analyzed otherwise than in a truly integral fashion.

The same thing is true for other traits in our species that are not susceptible to direct measurement. Humans search for meaning in music, art, works of charity, and the pure search for truth and beauty that is not related to economic gain.

In succeeding chapters, we will explore those traits and those yearnings. As we do, it will become increasingly evident that humans have many dimensions that have no equivalent in the animal species that preceded us.

One would think this is a truism – something so evident that the least educated in our species would easily acknowledge it, as was done by the young lady (Lauren) whom I quoted at the very beginning of this book.

And yet, it is often the most educated among our species who have a hard time acknowledging the importance of insights derived from the arts and from literature. When I was studying for my Masters in Public Policy at Harvard's Kennedy School of Government, I became restless when issues of war, as well as allocation of welfare by governments, were analyzed in totally quantitative terms.

One day, I couldn't stand it any more – the lack of humanity in the materialistic analysis of war and peace. It was an era in which, among the less erudite (but perhaps more empathetic members of our species), a book that was ostensibly written as a chidren's fable became popular.

The author's name was Antoine de St. Exupery. He wrote about a very wise, little prince.

The Lessons of "The Little Prince"

Unlike the technocrats at the Kennedy School, the greatest writers – and especially poets – never had to change their terminology. Since time immemorial, they have echoed Antoine de St. Exupery when he said, in *The Little Prince*, that "it is with the heart that one sees rightly; what is essential is invisible to the eye."

Poets, like theologians, tend to exaggerate. But the point made by the Little Prince stands: What is essential includes the material, the emotional and the spiritual.

Human happiness comes as much from what we aptly call the "humanities": art, literature, drama, poetry. And a lot of what we know about ourselves comes from the experiences of people who have suffered much yet have not lost their sense of purpose.

One in particular comes to mind as we seek to understand man's longing for meaning.

His name was Victor Frankl.

CHAPTER VI
MAN'S SEARCH FOR MEANING

Man's search for meaning is the primary motivation in [my] life. Victor Frankl

*V*ictor Frankl was one of the most celebrated victims of the Nazi Holocaust. This was in part because he survived – where so many millions did not. More importantly, it is because he managed to sew his memoirs into his prison garb and to later publish a book with the title used for this chapter.

Frankl was a respected psychiatrist and it would be hard for the psychiatry establishment to discard his dictum, even though it was expressed during circumstances that most of us never have to experience. Having survived the Holocaust, Frankl was in a special position to opine on how much our species derives meaning from values beyond the physical (what we today refer to as "creature comforts").

Here's a sample of his inspirational wisdom – of the higher values that were needed to survive the horrors of the Nazi concentration camps:

> *Woe to him who saw no more sense in his life, no aim, no purpose, and therefore no point in carrying on. He was soon lost...We had to teach the despairing men that it did not really matter what we expected from life, but rather what life expected from us.*

The West learned a lot from defeating the Nazis; the East, and in particular the Japanese, learned a lot from being on the wrong side of history.

Thereafter, and for a very traumatic half century, the West had to struggle against an evil empire that used the allure of material equality to suppress man's yearning for individual freedom. It was called "communism."

It was not until 1989 that the Soviet "Iron Curtain" began to crumble, and with it the idea that materialism was the only value worth pursuing.

Few writers, commentators or opinion-makers (with the notable exception of religious leaders and some politicians) seem to fully understand the danger that unbridled materialism brings. For reasons that I cannot comprehend, they don't grasp the macabre reality of the past century, when two materialistic philosophies (Nazism and communism) came close to closing their malevolent grip on humanity.

In my case, the historical reality is clear; I lived communism as a child, saw my father and sisters imprisoned without any judicial process, and was forced to live under house arrest. That was Cuba in 1961, as Fidel Castro tightened the noose on what was an island paradise and made it instead into a Caribbean version of the Gulag Archipielago.

What I don't understand is why Nazism and communism flourished in the very last century – just when planet earth was enjoying the fruits of liberty and the remarkable prosperity that came with the Renaissance, the U.S. and French revolutions and the industrial revolution.

In the case of communism, the book to read is *Witness*, by Whittaker Chambers, which describes the way in which the *intelligentsia* in America allowed the wool to be pulled over their eyes.

In the case of Nazism, the malignant role played by thinkers, including philosophers and ethicists, was astounding. By the time Hitler came to power, there were philosophical justifications for eugenics, euthanasia and ethnic cleansing. And they all stemmed from the same malignant, materialistic philosophy.

We will discuss those later in this chapter.

As for Nazism, everyone agrees that it represented one of the most frightening detours in the history of civilization. Yet there is little agreement as to why, in an age so advanced materially and politically, this awful regression by our species occurred.

Why, one wonders, did this awful ideological detour by our species happen to a society otherwise so advanced? Why this barbaric regression to pre-Judeo-Christian times, when in most areas of the world "might made right"?

I think it helps to look back at the values of that pre-Moses era – the ones we associate with Hammurabi's Code.

Pre-Christian Codes: From Hammurabi to Noah

Most historians and commentators refer to Hammurabi's Code as the key normative code of the pre-Jewish era. They cite its prescription of "an eye for an eye, a tooth for a tooth" as an example of what was a rather primitive, unforgiving way of life. Yet it was always made to sound equitable – albeit in a primitive, "pound-of-flesh" way.

It was much worse than that. In reality, Hammurabi's Code was quite hierarchical, cruel and discriminatory towards all classes of people, including women, slaves, and just about everybody who was not in the ruling class.

Before the Christian era – and outside of the Hebrew civilization – the accepted code of conduct was rather harsh. We forget just how radical the teachings of Abraham and Jesus were. Every code of conduct before that moment in history was far from egalitarian.

To understand fully just how much of progress was made in the establishment of a just society during the five millennia of Judeo-Christianity, we must look back a bit to the era before Abraham, when Persian and Assyrian civilizations thrived.

To hear many academicians today, you would believe that these early civilizations evolved, in linear fashion, towards the advanced culture and civilized behavior that we now enjoy. It all happened by chance, they argue, in the area known as the "Fertile Crescent."

Fertile Crescent As the Cradle of Civilization

The idea of a single, personal, omnipotent, loving God was a Jewish creation, later adopted by Christianity. It was bred in the Middle East – precisely in the area that Jared Diamond (and his modern followers) call the "Fertile Crescent," and to which they attribute the bulk of modern civilization.

I don't disagree with that theory, but only in the *geographical sense.* The most advanced civilizations were indeed bred in the Middle East and nurtured in the entire Mediterranean basin. But the cause was not, as Diamond argues in his celebrated work (*Guns, Germs and Steel*), the variety of plant species and weather systems that existed there.

Diamond's thesis leads him to suggest the most untenable theories. He is so convinced that geography and zoology control human development that he offers this totally unsupported proposition:

> *If the Old and the New Worlds had each been rotated ninety degrees about their axes, the spread of crops and domestic animals would have been slower in the Old World, faster in the New World. The rates of rise of civilizations would have been correspondingly different. Who knows whether that difference would have sufficed to let Montezuma or Atahualpa invade Europe, despite their lack of horses?*

It is inconceivable that this kind of stuff passes for science. The fact of the matter is that civilizations sprang in the Middle East and Asia Minor because of the confluence of longevity (they were probably the oldest humans), communications (they were the central highway from East to West, as well as North to South) and faith – which allowed neighboring tribes to trust one another, trade with one another, and learn from one another.

In other words, the Middle East was the historical epicenter of humans engaging in collective learning. The earliest written languages are found there. The most sophisticated structures and architecture were found there.

But the key ingredient that oiled human relations, permitting exchange of ideas and inventions, was faith. By that I don't mean a common religion; I mean faith in a loving creator, and consequently, faith in the basic equality of all humans, in the fundamental goodness of life on this earth, and in the idea that there are enough material goods to go around and that we don't need to fight for them, or deprive the weakest of our species of their chance to pursue happiness.

Those ideas of equality, as we shall see, blossomed and prevailed in the Mediterranean basin and spread north to Europe, allowing it to flourish. Had they not, Europe would not have been the cradle of intellectual enlightenment, industrial revolution and the bulk of modern science – let alone the kind of egalitarianism that accompanies democracy.

Northern Europe, in particular, was able to flourish primarily because it adopted the Judeo-Christian ethic and ethos. Had it not, we might have continuing doubts as to its ability to surpass its barbarian roots. Those were so primitive, so cruel, so based on raw power that Aristotle actually expressed doubt that Northern Europeans would ever be able to govern themselves.

In summary, we accept the geographic genesis (the "Fertile Crescent") of what we now call "European" or "Western" civilization. A better term, geographically speaking, would be "Mediterranean" civilization, since the crescent includes the Middle East, Asia Minor and North Africa.

Yet we accept very little else of Diamond's analysis. The truth is that it wasn't just geography, geology, or botanical diversity that forged civilized behavior and prosperity. Instead, as argued by Ben Shapiro, it was the wisdom and the faith passed down by Athens and Jerusalem.

Of Athens and Jerusalem

Shapiro concentrates on Athens and Jerusalem, but the cradle of civilization is also very much Rome – and for that matter, the areas we now know as Turkey, Syria, Lebanon, Egypt and Carthage. In other words, the Middle East and both shores of the Mediterranean.

But, again, geography is not the unifying factor; nor is geography the propellant of civilization. The moving and unifying factor of great civilizations is ideas. And many, if not most, of those ideas, are faith-inspired.

Here's how Shapiro explains it:

> [I]*deas matter, and important ideas - as best articulated by great thinkers – represent the motivational road along which humanity journeys. We act because we believe.*

How true!

And the ideas to which Shapiro refers – the ones that represent the "motivational road" along which humanity has journeyed since about four thousand years ago, is the Judeo-Christian code of ethics.

How Two Roads Diverged

Any impartial, comprehensive reading of history reveals that peoples separated culturally and geographically from Jerusalem and Rome, as the binodal cradles of Judeo-Christian civilization, had a much slower path to egalitarian conduct than Jews and Christians. In fact, before Abraham and Jesus, most societies divided their members into "strong" and "weak."

And almost everyone was in the latter category. Our species, before the advent of Judeo-Christianity, had always classified the weaker members as inferior. Based on biological strength alone, women were always characterized as the "weaker sex."

That, of course, is a term no longer used, in part because we realize it's discriminatory and in part because we realize it's dead wrong – even in the strictly biological sense.

By almost any definition of strength that we can objectively consider, women are as strong as men. They bear pain much better; endure sleeplessness where most men would be reduced to whimpering and giving up, such as when nurturing a sick child for 24-hour or 48-hour cycles. They are at least as hardy when it comes to enduring extreme heat or extreme cold.

Physiologically, a good argument can be made that women can handle the stress of life much better than men; they certainly live longer. (The maximum age for women is two years longer than for men; the age expectancy shows an even greater disparity; in the U.S. it is about five years.)

Yet in antiquity, most societies went by the principle that "might makes right," and, in general, men had superior physical strength. So they subdued the women and treated them as inferior beings.

Explorers like Marco Polo described the way women were treated well into the thirteenth century in the Asian continent, when it was ruled by Ghengiz Khan and his successors.

The ruler, of course, had his choice of wives. The power over women that rulers had was made evident by this scene in Marco Polo's narratives, in which he explains what happened when the Khan of Persia "lost his favorite wife and desired that another be sent to him from the same Mongol tribe from which she had come." Marco Polo and his delegation were happy to oblige, but here's what happened on the journey:

> *It took them more than two years to reach Persia where, alas, the Khan was already dead and the bride was delivered to his son instead.*

I wonder if the young bride was happy that her intended husband was dead and that she was given over to the younger ruler.

Either way, what is clear is that women were little more than chattel.

And it remained so until Western civilization reached Asia and the Far East.

It didn't happen naturally, and it didn't happen by natural selection based on random mutations resulting from the effort to survive. Instead, it happened, in great part, from the proselytizing efforts of missionaries (most of them celibates) and other well-meaning souls who were inspired by a sense of brotherhood of all men and women.

Without that influence, the indigenous people might have never "evolved" at all in their habits, their subservience of women and children, and their insouciance towards those afflicted by disease or age.

Occasionally, some who profess to use the tools of science, have argued otherwise.

The Truth About Indigenous Peoples

In the middle of the last century, a number of well-meaning scientists (including, and most prominently, Margaret Mead) spent some time with aborigines and tried to convince the world that isolated tribes were kinder and gentler towards one another than Western civilization.

Those ideas have mostly been debunked. I don't know of any respectable scientist, nowadays, who considers aboriginal societies to be kind or gentle – particularly towards the young, the old and the women.

Here's an illustrative passage from Diamond which describes some of the traits of tribes in an isolated region that we now call New Guinea:

Forms of self-mutilation and cannibalism varied from tribe to tribe. Child rearing practices ranged from extreme permissiveness (including freedom for babies to grab hot objects and burn themselves), through punishment of misbehavior by rubbing a Baham child's face with stinging nettles, to extreme repression resulting in Kukukuku child suicide. Barua men pursued institutionalized bisexuality by living with the young boys in a large communal homosexual house, while each man had a small heterosexual house for his wife and daughters and infant sons. Tudawhes instead had two-story houses in which women, infants, unmarried girls, and pigs lived in the lower story while men and unmarried boys lived in the upper story accessed by a separate ladder from the ground.

Clearly, the areas of the world that did not experience the Judeo-Christian code of ethics did not experience the same civilizing influence that the Western world did.

Some anthropologists tried to resist this simple historical fact, but there is no denying it. They try to romanticize what was happening in Asia during the last two millennia, but there is no denying that life in Asia was barbaric – which is what you expect when barbarians rule the roost. Yet look at the facts.

Here's a statistic from Jared Diamond himself on what was happening in central Asia for the eight centuries between Attila the Hun (434) and Genghis Khan (1262), when as he describes it, barbarian hordes "overran the area." By way of quantitative measure, he states that "scholars debate whether Genghis Khan's armies slaughtered 2,400,000 or only 1,600,000 persons."

Genghis Khan was not the last of the invaders from the East. Turks waged war on Europe until the battle of Vienna (1683) and Moors wreaked havoc in the Mediterranean until the United States created a navy and force-fed peace treaties in the nineteenth century. (The book to read is by Michael Oren, *Power, Faith and Fantasy*.)

But this happened much later in history. To get there, humanity had to first experience 40 centuries of Judeo-Christian civilization. Let's backtrack a little bit and review that history.

The Noahide Laws

It was not until the era of Noah, approximately 4,000 years ago, that a reasonably humane societal code emerged; it contained seven laws and is referred to by the title of the "Noahide Laws." Shapiro summarizes the seven laws as "designed to minimize human cruelty" and consisting of "bans on murder, theft, idolatry, sexual immorality, animal cruelty, cursing God and the positive commandment to set up courts of law in order to punish crimes."

Take out the two theological tenets (banning idolatry and cursing God) and you pretty much have the basic system of justice which Aristotle and Aquinas baptized as "commutative justice." The term "commutative" really means "reciprocal." In other words, respect the property and person of your fellow beings – and abide by contracts freely entered into.

In summary, the normative fabric of civilization was woven by Judeo-Christian ideas, which took root in Diamond's Fertile Crescent. Those ideas spread from Jerusalem, Athens and Rome to the far corners of Northern Europe and North Africa, propelled by the Roman empire.

Athens is important because so many of the seminal ideas of justice come from ancient Greece, whose thinkers gave us the concept of a natural code of conduct.

Cicero (106 B.C. to 43 B.C.) was particularly clear on the idea that our reasoning mind could ascertain that code or law:

> *True law is correct reason congruent with nature, spread among all persons, constant, everlasting. It calls to duty by ordering; it deters from mischief by forbidding....It is not holy to circumvent this law, nor is it permitted to modify any part of it, nor can it be entirely repealed....*

Cicero was also clear that the law dictated by a correct view of nature drew its validity from the belief in a creator. Here's how he explained it:

And one god will be the author, umpire and provider of this law. The person who does not obey it will flee from himself and defying human nature, he will suffer the greatest penalties by this very fact, even if he escapes other things that are thought to be punishments.

The best compilation of a natural code of conduct goes back a few centuries before Cicero – to the time of Moses (circa 14 century BCE) and the Dcalogue, or – as we generally call it today – the "Ten Commandments." Historian Thomas Cahill refers to that religious and cultural tradition as the "Gifts of the Jews."

The Gifts of the Jews

The intellectual seeds of civilized behavior were sown by a small tribe known as the Hebrews. Many great historians have written about that in words more eloquent than I. One of those is Thomas Cahill, who waxed particularly eloquent on the impact of the Ten Commandments, or Decalogue.

Here's Cahill:

If I can peer through the mists of history and see the begrimed, straightforward faces staring upwards towards the terrors of Mt. Sinai and if I can imagine the immense throng of simple souls trudging through the whole of history – all the ordinary people down the ages in need of moral guidance in all the incredibly various cultures and situations that the planet has known, it must be admitted that it would be fairly impossible to improve on the Decalogue as we have it. The sins it catalogues are the great sins and those it does not mention explicitly – such as withholding sustenance from those who have nothing – can be deduced from it.

Cahill's prose may be a bit flowery, but you get the point. The arrival of the Hebrew civilization, with its Ten Commandments and its notion of a single, almighty, righteous God was a key element in civilized conduct in the history of the Middle East. And the Middle East was the epicenter of culture and trade. The Fertile Crescent was truly fertile – but not in the way Diamond meant it. It was fertile in terms of the Judeo-Christian *ideas and ideals* that spread north and south.

It also became the epicenter for enormous and fortuitous *spiritual fertility*. With the birth of Jesus, our species experienced a totally unexpected, unprecedented and epochal change in the way humans viewed one another. Whether it was inspired by God or not, it was a magnificent story.

It's been called "the greatest story ever told."

The Greatest Story Ever Told

The accelerated path of compassion and brotherhood came with the arrival of Jewish carpenter called Jesus. In a societal sense, it was not so much the birth of Jesus, nor the subsequent expansion of Christianity to the non-Jewish world, as narrated in the "Acts of the Apostles." It really began with the conversion of Emperor Constantine (312), when the followers of Jesus went from being persecuted pariahs to being wholesale transformers of the socio-political milieu.

Here's a passage from the recent (2018) publication by scholar Bart D. Ehrman:

> *Christianity not only took over an empire, it radically altered the lives of those living in it. It opened the door to public policies and institutions to tend to the poor, the weak, the sick and the outcast as deserving members of society. It was a revolution that affected government practices, legislation, art, literature music, philosophy, and – on the even more fundamental level – the very understanding of billions of people of what it means to be human.*

It was a whole new way of thinking; a revolution, as Ehrman describes it, which affected laws, art, literature, music, philosophy and, as per the last phrase above, our "very understandi of what it means to be human."

Thankfully for our species, it soon spread to the entire Western world, mirroring the expansion of the Roman empire.

The Roman Empire Expands

Within four hundred years after Jesus, the Roman Empire had become Christian. And by the end of the millennium, all of Europe had. Eastern Europe – more specifically Hungary – became Christian when the last Magyar accepted a kingly crown from the pope and became Stephan I of Hungary (1,000).

Thereafter, most historians separate the planet into West and East. For the next millennium, and until well into the nineteenth century, Europe and American colonies of Europe (the so-called "New World") thrived while the East struggled with atavistic customs, tyrannical systems of government and delayed industrialization.

What distinguishes Western civilization from the others that prevailed, until very recently, in other parts of the world? What puts Europe and North America on the "right side" of history?

These questions constitute the theme of the previously mentioned book by Ben Shapiro, whose title (*The Right Side of History*) encapsulates the victory of Western civilization, with its liberal ideas of human rights and of rule by the majority – what we call democracy.

But let's not get ahead of ourselves. From the moment of conversion of Constantine, in the fourth century, to the Middle Ages and the Renaissance that followed, there is a period of history which is referred to as the "Dark Ages."

Let's analyze what was happening in Europe during those eight centuries and why it is historically imprecise, if not downright erroneous, to call that era the "Dark Ages."

The Not-So-Dark Ages

History is made easier today by the computer. We can confirm most major historical events by analyzing and comparing enormous volumes of newspaper accounts, personal letters, and all kinds of documents that have been unearthed, catalogued and shared by scholars with ease.

By all accounts (except those that are clearly tinged by ideological bias), the eight centuries after Constantine embraced Christianity were not a historical wasteland of obscurantism or wholesale regression in the expansion of civilization and culture.

What is true is that Europe, with the fall of the Roman empire, experienced renewed attacks by uncivilized barbarians from the north, east and south. From the north it was the various Celtic, Gothic and Nordic tribes. From the east it was the Magyars and the Huns. From the south it was the Moors.

Europe was besieged, during the better part of a millennium (400-1200), by barbarians. Yet Europe not only survived the onslaught, but managed to prosper and grow in every aspect of cultural life, from literature to technology and the plastic arts.

Clerics from Ireland populated northern Europe and brought with them classical scholarship, including Hebrew and Arabic texts, to which they painstakingly added the Christian bible and its literary offshoots. Western civilization arose from the combination of those Middle Eastern cultures, which were preserved in all parts of Europe – not just northern Europe.

Special mention should be made to the period when southern Spain was experiencing an era of "Convivencia," when Arabic, Hebrew and Christian cultures not only coexisted, but thrived.

By the thirteenth century, Thomas Aquinas and the Scholastics had blended all of those cultural strains into their writings. And I should clarify that the Western values that they embodied is not in any way a reference to the predominantly Caucasian folks who ultimatelty became the predominant ethnic archetype of Europe. In fact, the nations (tribes, really) that were conquered or repelled in the northern regions were as barbaric as the Asian ones from the east or the African ones from the south.

As previously mentioned, Hungary was ruled by barbaric Magyars until the conversion to Christianity of Stephan I in the year 1,000. The Moors, coming into Spain from the south, were ultimately defeated by Pelayo of Asturias (c.718-737). As to the Celts and the other tribes of Northern Europe, they were converted as much by the good example of Christian monks as by the sword.

By the beginning of the second millennium, Europe had become the seat of philosophy, liberal arts, science and the Judeo-Christian normative code. Thereafter and for another millennium, civilization thrived in Europe. Ideas hatched in the Mediterranean and Northern Africa bore fruit in France, Spain, Italy, Germany, as far east as Russia, and as far north as the British Isles and Scandinavia.

All the time, Irish monks were preserving the writings of both Jewish and Arab thinkers.

While the rest of the world was destroying the great books, clerics in Europe were preserving them. They were also forging a philosophical fusion of all the best ideas of east and west.

Contrary to what Aristotle predicted, the Northern European tribes not only learned to govern themselves, but became the new epicenter of civilization and democratization. Ideas born in Athens and Jerusalem spread via *Pax Romana* to all corners of the Western world.

It should not be forgotten that the Roman empire extended south, to Africa, as well as north. In fact, one of the great scholars – arguably one of the three greatest philosophers of all time (together with Aristotle and Aquinas, in what I call the "A-Team") – made his home in North Africa.

He was known as Augustine of Hippo.

St. Augustine and the Search for Truth

Augustine of Hippo (modern day Tunisia) was the first thinker who suggested that humans are simultaneously citizens of a heavenly city as well as an earthly city. Humans, to be fully fulfilled, must follow both a divine law and a human (or natural) law.

Nowadays, we struggle a lot with the concept of the supposed tension between "church" and "state." To Augustine there was no tension whatsoever, for the simple reason that one God created both. To that end, he once famously said: "Intellege ut credas; crede ut intellegas." Reason so you may believe; believe so that you may reason.

Reason and faith go hand in hand in the search for truth and human happiness.

It sounds simple enough, but it is not so simple, as a matter of practice. For most of modern history, most societies have proclaimed either official state religion or an official state secularism. One can argue about which has been worse for humanity. As previously stated, both of the evils that plagued the last century (Nazism and communism) were premised on the idea that man is an expendable commodity with no God-given dignity or inalienable rights *per se*.

In the case of democracy, every political scientist I have read has argued that in order to prosper a democratic society must have individual members who have internalized a behavior code that calls for tolerance, respect for the property of others, and respect for the family of others.

It helps a lot if they also believe that there is a higher purpose to life than mere survival of the fittest, or even survival of their own tribe. As I explained before, the belief that all members of our species deserve a dignified life was born in the Middle East and took root in Europe; thereafter, the idea expanded to all other continents, including the New World of North and South America.

It led, inexorably, to a more complete form of justice. But, first, let's analyze the conceptual development of what has come to be called the "natural law."

Building on the teachings of Plato and Aristotle, Thomas Aquinas and his Scholastics fused together the great ideas of humanity, going back to the Greeks and including the moral lessons of Judeo-Christianity. Here's how Ben Shapiro describes what he called this "grandest attempt to reunify Athens and Jerusalem":

> *During the twelfth century, Aristotle's works, long buried, were rediscovered in the West. They had been maintained in the Arabic-speaking world for generations, but they were only retransmitted in Europe over the course of that century, breaking anew and with massive impact in the thirteenth century.*

The Scholastics fused the tenets of the natural law, arguably derived from reason alone, with the teachings of Jesus, which admitted of some flaw in our natures that required some form of "redemption" or what modern psychologists call "behavior modification." These thinkers took note of a reality – that humans occasionally follow their baser instincts, and that these instincts often overcome the precepts of the natural law, as are evident from reason and experience.

Divesting the human capacity to reason from the bouts of irrationality that are impelled by our instincts gone wild, they came up with a slightly more advanced idea, which they called "right reason." In other words, this is the capacity of humans to exert control over what neuroscientist Dingman – cited in Chapter III above – refers to as "pleasure-seeking/pain-avoiding stuff" in the human brain.

The Scholastics built on the wisdom of the Hebrews and the Greek savants; they insisted that all humans have equal dignity and concomitant rights. They added the notion of equality to the natural law. But it was not just egalitarianism of the kind that treats all equally before the law, period. It prescribed a greater level of equality that was based on the idea that all humans were thought to be created in the "image and likeness" of the creator, and deserved a dignified lifestyle.

The two elements above – right reason and equality – led to what ultintely became a universally accepted notion of the natural law. It was defined by Hugo Grotius (1583-1645) as the –

> ...*dictate of right reason which points out that an act, according as it is or is not in conformity with rational nature, has in it a quality of moral baseness or moral necessity...*

It was the beginning of the era of "renaissance" and accompanying "enlightenment." By this time, other great thinkers chimed in, applying dictates on the natural law to economic arrangements. One such thinker was the father of economics, Adam Smith (author of *The Wealth of Nations)*,who argued for economic ("natural") liberty within the bounds of what he called the "laws of justice." Here's how he defined "natural liberty":

> *The obvious and simple system of natural liberty establishes itself of its own accord. Every man, as long as he does not violate the laws of justice, is left perfectly free to pursue his own interest his own way, and to bring both his industry and capital into competition with those of any other man, or order of men.*

During the next few centuries, the Western world struggled with the balance between individual liberties and collective welfare. "Conservatives" and "liberals" (also referred to as the "right" and the "left" of the ideological spectrum, respectively) battled over how much equality could be imposed by law, how much right to govern was vested in kings, and how much individual choice could be allowed before society evolved into pure anarchy.

Moreover, the evolution of individual rights continued in the next three centuries, merging into the collective human aspiration to have a say in how societies are governed. This aspiration flowered in a special way in the New World.

In the eighteenth century, American leaders like George Washington, Thomas Jefferson and Alexander Hamilton established an independent and totally democratic, continent-sized nation.

In the nineteenth century, leaders like Abraham Lincoln and Simon Bolivar showed that humans of all races and creeds had the right – as well as the competence – for self-government.

By the twentieth century, leaders like Mahatma ("great soul") Gandhi completed the task of giving all peoples of the world the right to self-government.

Unfortunately and inexplicably, the twentieth century also brought with it two deadly collectivist plagues, each of which almost destroyed our species. Humanity experienced the collectivist extreme, in not one, but two manifestations.

Our species learned to reject both Nazism and communism.

Extreme collectivist regimes violate the right of individuals to produce and consume as they see fit. The opposite extreme (Libertarianism) refuses to recognize the rights of all to a fair share of the resources of the earth or of the production that is, in great part, made possible by society as a whole.

Let's discuss that extreme and its philosophical ramifications.

Extreme Individualism: Ayn Rand and the Libertarians

Commutative justice by itself is not enough. The reason is simple: Many members of our species are at a disadvantage at some point in their lives. Right away, children and the elderly come to mind. In the Judeo-Christian scheme, those must receive special protection because they are inherently vulnerable.

Examples of that special protection include criminal laws prohibiting sexual contact with children and civil laws allowing elderly citizens to void contracts for a period of time (referred to as a "cooling off" period) after they sign them.

The same compassionate protection is extended to those born with handicaps – now referred to as having "special needs," as one of my eleven grandchildren has.

Early on in the history of civilization (approximately 5,000 years ago in the case of the Jews), enlightened members of our species realized that treating all persons as equal before the law might not be enough for a just society. People make mistakes. Some become addicted to alcohol or drugs without realizing it. Others have a hard time concentrating or learning enough to be self-sufficient.

Western society, based as it was (and still is, to a great extent) on free markets, provided an excellent platform for the community as a whole to prosper based on the selfish choices of individuals to consume and produce. The theory, advanced brilliantly by Adam Smith, was referred to as the "invisible hand" of the market, which allowed supply and demand to determine spontaneously the ideal price for goods and services.

The extreme view of that theory, called *laissez-faire* capitalism or libertarianism, says that we should never interfere with free-market competition. Companies and individuals that go bankrupt deserve to go bankrupt; those that cannot get a high-paying job should do menial work and eke out a living.

Humans, they argue, are just like animals in that sense. It's dog-eat-dog out there and we should not interfere with that natural balance. It is very much like Darwinism, applied to the human society; hence it is often referred to as "social Darwinism."

Social Darwinism, combined with strong doses of racism, eugenics and nationalism laid the intellectual foundations of Nazism. Suggesting that the Germans were at the top of the evolutionary heap, this dark, materialistic and cruel vision of humanity threatened to destroy civilization.

Its principal philosophical exponent was Friedrich Nietzsche. Admired by many for other reasons, Nietzsche wittingly or unwittingly articulated a justification for social Darwinism, and from there it was an easy jump to Nazism. Yet there was no justification. It was godless; it was materialistic;

it was cruel. And when implemented in Nazi Germany, it almost led to the destruction of all modern humanistic societies.

Let's analyze its theoretical underpinnings.

Nietzsche and the Rise of Social Darwinism

During the early part of the last century, a school of philosophers, led by Friedrich Nietzsche, tried to take God out of the equation, ending with the macabre belief in social Darwinism.

Here's how historian Will Durant describes it:

If life is a struggle for existence in which the fittest survive, then strength is the ultimate virtue and weakness the only fault. 'Good' is that which survives, which wins; 'bad' is that which gives way and fails....Men who could think clearly soon perceived what the profoundest minds had known: that in this battle we call life, what we need is not humility, but pride, not altruism, but resolute intelligence; that equality and democracy are against the grain of selection and survival; that not masses but geniuses are the goal of evolution; that not 'justice' but power is the arbiter of all differences and all destinies. So it seemed to Friedrich Nietzsche.

Nazism took root slowly but surely by playing upon the national pride of Germans, at the emotional level, and on the social Darwinism of Nietszsche, at the intellectual level.

It was also a reaction to the other materialistic and collectivist philosophy of the times, which was the one that gave rise to communism, buttressed by the writings of Karl Marx.

The twentieth century was thus convulsed by two materialistic, godless philosophies and their aggressive adherents – the Soviet Union and Nazi Germany.

By the time the Second World War was coming to an end, America's president, Franklin Roosevelt (FDR), having defeated Nazism by the sword, was looking to defeat communism by the word.

FDR was aware of the allure of communism, with its vision of a society where equality was imposed by the state, even at the expense of individual liberties. He was looking for a middle ground between the unrestrained individualism of free-market bankers and industrial monopolies (whose excesses had been greatly responsible for the Great Depression) and the coercive, collective materialism of Soviet-style socialism.

Thankfully, he was inspired by a young aide, and later by the first woman cabinet member.

FDR's "Freedom from Want"

In his eminent biography of FDR, Jonathan Alter discovers for us how the seeds of social justice were sown for the young, future president. Alter narrates a scene in which a young associate of Franklin, named Ed Walsh, said to him: "Franklin, you don't know much about political philosophy; you should start by reading this." And "this" was a papal encyclical that later became famous; it was written by Leo XIII and contained the basic principles of workers' rights.

Pope Leo XIII's and FDR's fusion of the two kinds of justice gave humanity a rational manual for the conduct of not only the individual, but the society of individuals. Social justice would henceforth guide the dealings of groups of humans, just as commutative justice guided the fair conduct of individual transactions.

As previously stated, the concept of distributive justice goes nowadays by the name "social justice." Social justice is a concept that embraces two aspects of the human, economic condition: wealth and income.

Often, as we saw in the the development of the American west, as well as the industrial east, individuals accumulated enormous wealth. First under Teddy Roosevelt and later under Franklin Roosevelt, a concerted effort was made to divide the wealth, lest certain industries become monopolies that could fix prices, as well as elections.

On the income side, the idea that no one should be left behind, in a society as rich as those of Western Europe and the United States, began to flourish with the aforementioned Papal encyclical. Written in the late nineteenth century, this manifesto on social justice scandalized some conservative Catholics, who were sure that the right to the fruits of one's labor was unrestricted.

But the reformers would not rest. Led by certain Protestant sects, such as the American Episcopalians, the western world began to reach the conclusion that all people should share in the bounty that nature, coupled with human ingenuity, provides.

A half century later, as WWII was ending, it fell to FDR to proclaim the basic principle that all persons are entitled to have a decent standard of living; curiously, he did it as a double-negative, perhaps persuaded that proclaiming a positive human right was too radical. He called it "Freedom from Want."

It is said that the person who suggested the formulation of the fourth great freedom was FDR's Secretary of Labor – and the first woman member of the cabinet – Frances Perkins.

It is also important to note that distributive justice derives support not only from philosophy but also from economic theory. By mid-twentieth century, economists were developing models that tended to support limitations to unbridled, enlightened self-interest.

It's instructive to see how economics and philosophy converged to advance the cause of what we now call the "common good."

Economists Define Collective Welfare

While Adam Smith proved that the forces of supply and demand generally reach a desirable equilibrium if left to their own devices, in practice the economic model needs a number of constraints.

To begin with, there are many "goods" which the market will not produce on its own, because they require enormous up-front investment. Classic infrastructure like highways and railroad tracks require an up-front, collective effort that only governments are willing and able to undertake.

Another constraint on the free market is its own tendency to monopolize a particular resource or industry. In the United States, many industrial giants had to be broken up in order to maintain competition and avoid price-fixing.

The use and ownership of other resources, such as beaches, fresh water and certain scarce minerals also must be regulated to prevent private individuals from appropriating them to the exclusion of the average citizen.

Last but not least is the situation where one party has unequal bargaining power. The extreme case is the "natural monopoly," which is an industry sector whose ideal size is an entire region. Possessed of that sector, which can be – for example – a gold mine, a port or an airport, the private-sector operator can reduce wages to what is called the "marginal" value of one additional employee.

To equalize the bargaining power, employees must bargain collectively by forming labor unions that force the owner to compensate employees on the *average* value of their work, rather than the *marginal* value.

Those, in summary form, are economic reasons why a society cannot function under the banner of extreme individualism. There are others, including the environmental interdepencies of our use of resources. (Ecologists call those "externalities.")

But here I want to discuss the need for compassion and sharing that makes any society richer as a whole, while taxing individuals.

The philosophical principle is known as the "common good." It sounds to some like communism, but it is far from that, because it recognizes the infinite value of the individual. It balances the collective good with individual human rights. And thus it strikes a perfect balance between extreme individualism and extreme collectivism.

The above are technical considerations, but they play nicely into principles that emanate from philosophy, sociology and psychology. All the natural and social sciences converge in the notion that our species is entitled to both individual rights and also some measure of social justice.

Theologically, it became a fusion of the Ten Commandments and the Beatitudes – of the inviolability of the person but also the welfare of the community. This balance was increasingly achieved in the western world.

The Idea of a Common Good

Let's summarize.

Extreme individualism leads to a society in which those who are physically weaker, or fall prey to addictive substances, or have unequal bargaining power to obtain a fair wage will either not survive or not be treated with dignity by their fellow beings. In the cradle of civilization which was the Middle East (Jared Diamond's "fertile crescent"), the prevailing rule of law was the inverse of the Golden Rule.

The colloquial inverse of the Golden Rule goes humoristically like this: "He who has the gold rules." We laugh now, but up until the modern era, in most areas except the area that embraced the Judeo-Christian code of ethics, the history of our species was precisely that.

Everywhere in the ancient world, except for areas where the Ten Commandments and the Beatitudes prevailed, it was "might makes right."

Judeo-Christianity turned that hierarchy upside down. Buttressed by Old Testament prophets, including Amos, Jesus preached that the person who has received much is tasked to share much. That the strong must serve the weak and the rich must share with the poor.

Western society made that into a code that exalts the common good, without reducing the individual or making him subservient to the state. Modern societies, organized in that mold, subscribe to the principle that no member of our species should "fall through the cracks." No human should be deprived of his or her basic needs.

Winston Churchill referred to that concept as "competition upwards, not downwards." Among other reforms, he advocated for universal right to health care. Soon, all western democracies adopted that as either a goal or a right.

The Western Democracies Adopt Social Justice

By the beginning of the last century, the Western democracies, including most of Europe and North America, had reached a consensus that called for some basic guarantees for all their citizens. Among the things guaranteed were elderly benefits (Social Security in the U.S.), jobless compensation and a measure of welfare for the poor, including foodstamps.

Following the lead of England's Winston Churchill, who had advocated for universal healthcare, the United States enacted legislation that provided health care to the elderly received in the form of Medicare and to the poor in the form of Medicaid. In America, food stamps and affordable housing were provided for a big percentage of the elderly and poor. And public education, from kindergarten to 12th grade, was provided at no cost to all.

It was not a welfare state *per se*; most of the Western economies had a strong free-market component, in which companies and individuals risked capital to make a bigger profit. The great majority of political leaders did not want to eliminate competition, which they recognized was a motivating force of inventiveness and the fulfillment of the human desire to exercise individual economic freedom.

Moreover, it was the only form of government that worked.

But deep in the culture of the developed nations was also a recognition of the ultimate dignity of all members of the species. Charitable and civic organizations filled a lot of the needs of the poor, the elderly and the sick. Christian and Jewish schools taught the children – often with scholarships for those who could not pay. (The book to read is *Democracy in America*, by Alexis de Tocqueville.)

Ultimately, primary education became free for all and near-free up through a two-year community college. Contrary to Jesus's statement that "the poor will always be with you," most of society accepted the notion that no one should ever go hungry or homeless or without medical care.

It was no longer a dog-eat-dog world.

How much of this egalitarianism was prompted by faith in an almighty being that rewarded generosity in an afterlife? Is it possible that humankind would have spontaneously continued formulating the kinds of norms that would have allowed the world to reach the kind of civilization we enjoy now? In other words, would humanity, without the Judeo-Christian inspiration, have ever reached the stage where every citizen is equal under the law and that guarantees every one a decent standard of living?

I leave it to my reader to reach his/her own conclusions. For me, it's evident: Much, or perhaps most, of the goodness in modern society follows from the Judeo-Christian inspiration. And this applies to democracy just as much as to other basic freedoms.

Would modernity, without that inspiration, ever have embraced the concept of one-person-one-vote? Or of universal health care? Or social security for the elderly? Or the right to *habeas corpus*, and ultimately the right to be tried by a jury of your peers?

It would take an enormous amount of sociological and historical research to determine if our species would have reached the current level of civilization using the tools of science alone. It says here, based mostly on anecdotal evidence and intuition, that without the Judeo-Christian inspiration, humanity would not be anywhere close to the present state of welfare, self-government and the sheer dignity that most societies vest on the individual.

Humankind might or might not have evolved, in the 5,000 years since Abraham, to the level of freedom and welfare enjoyed by most people today. I have already described the barbarism and inequality of the era before Abraham, when Persian and Assyrian civilizations thrived. I have analyzed the impact of the Scholastics, including their development of the natural law and advanced concepts of justice.

Much of the logic and many of the ideas have been rooted on reason.

Now let's review what happens when the element of faith is added to the great ideas that led to the European Renaissance and to the remarkable level of welfare and liberty that characterizes much of the modern world.

Faith is like the icing on the cake, where the cake is made of two components – reason and experience – whose proper interaction forms the basis of science.

In the next chapter, we delve into the spiritual dimension of our species from the standpoint of both empirical and social science. The objective is to see where science and religion interface.

CHAPTER VII
OF FAITH AND SCIENCE

This whole experience has taught me that it's not by accident,
it's by design. When you truly trust God….you know you're
here for something greater than yourself. Tracy Mourning

$\mathcal{T}$racy Mourning was, for almost two decades, the celebrated, adored companion of professional basketball star, Alonzo Mourning. Recently, her marriage ended in divorce.

Alonzo and Tracy Mourning are living human icons. They are both gorgeous, tall, svelte, and possessed of the creamy skin tone that appeals to all races and ethnicities in the modern world.

They are also immensely wealthy and popular. In many ways, they are Miami's king and queen, rivaled, in recent years, only by (former Mayor) Maurice and Mercedes Ferre, and arguably, at this particular moment in history, by the current mayor of Miami and his lovely first lady. (I am biased in that assessment, as that is my son, Francis and daughter-in-law, Gloria.)

But even these human icons meet failure and encounter human grief – though not necessarily of the material kind. Tracy Mourning will most likely always be rich beyond our comprehension. And she will surely be beautiful into the latest stages of her life.

Alonzo stands just about seven feet tall, with chiseled body and he face of an Egyptian pharaoh. He is the first professional athlete that I saw pointing his hand in the air after scoring a basket, as if to recognize the role that he believes the Almighty plays in his own athletic accomplishments.

The role that faith plays in the African-American community is particularly important. In one of his narratives describing a lengthy visit to the United States, the great Lutheran theologian, Dietrich Bonhoeffer, stated unequivocally that the expression of faith in black churches was more notable and much more emotionally poignant than in white churches.

Bonhoeffer was particularly impressed with the quality of the homilies, which he thought captured the biblical message much more vividly and profoundly that those in white churches. Perhaps it is a function of the same kind of faith that motivates people who have suffered much. In Miami's inner city, as in many others in the United States, people have suffered much.

Here's a vivid example of that faith.

The Christian faith is evident in a hunger strike that took place in Miami about a year ago. In the same television interview that dealt with Tracy and Alonzo Mourning, the commentator (Jawan Strader), also covered the 21-day hunger strike carried out by nine African-American men.

It is a story of faith like few that I have ever witnessed; and this one was firsthand.

The Hunger 9

In the television interview of the "Hunger 9," the theme of God comes up repeatedly. If there is no almighty being – if there is no being "greater than oneself" who has something more long-lasting in mind for the human species, then a lot of things that people say and do make no sense.

The hunger strikers were mostly ex-convicts who were concerned that the next generation of African-American youth would continue a historical pattern of gun violence. As a symbolic gesture, they went without food for 21 days.

As I watched them and spent time in their company, I realized that I was not capable of that sacrifice. I know for them there was some measure of notoriety, as they were put on billboards, feted by political leaders and had their life stories told in newspapers.

Still, there was an element of pure altruism. They really wanted kids not to follow their example. And they did something heroic.

The sociobiologists don't buy the idea of altruism, or rather that it is faith-based. They are convinced that somewhere along the line, group survival required that invididuals subsume their own welfare to the group. Altruism to them is just another survival trait, only a collective one.

The best known modern proponent of that theory is Richard Dawkins.

Richard Dawkins: "Selfish Gene"

In the cruel new world of sociobiology, nature (which includes our species) is seen as "red in tooth and claw." In other words, it's dog-eat-dog out there. The philosophical equivalent, as previously stated, is scientific materialism; its most fervent spokesperson is Richard Dawkins.

Dawkins is sort of a literary agent for the movement that seeks to convince us that humans are just slightly smarter animals. He's the loudest voice out there, but the pioneer and still best-known champion of sociobiology is Harvard's E. O. Wilson. Here's a sample of Wilson's wisdom, as described by Ben Shapiro:

> *E.O. Wilson...posited that human beings had inescapable programming that made us behave in certain ways in response to our environment. Furthermore, through investigation of the interaction of that innate nature and the environment, we could fully predict human behavior.*

As Wilson has done to expand sociobiology into the realm of psychology, Dawkins has done to expand it into philosophy. And so he spends his writing energies frontally attacking the notion of faith or of a higher-purpose for which our species could possibly devote itself.

He fancies himself a philosopher.

That is not the case for other sociobiologists; as far as I can tell, these folks simply ignore other dimensions of the human species. The result, as shown in previous chapters, is a colossal misunderstanding of the multi-faceted reality that constitutes our species.

Dawkins, as the classic scientific materialist, is not so constrained. He coined the term "selfish gene" to describe what he considers the defining, exclusive building block of our species. He is convinced that the human species, as a product of evolution, must be selfish to the core.

And yet even Dawkins, the convinced atheist, struggles with the communal implications of universally selfish conduct by the members of society. Notice how he struggles with the logic of his own science, which would dictate that our species is as "red in tooth and claw" as any species in the animal kingdom:

> *Be warned, that if you wish, as I do, to build a society in which individuals cooperate generously and unselfishly towards a common good, you can expect little help from biological nature. Let us try to teach generosity and altruism, because we are born selfish.*

That's quite a *non-sequitur*. Dawkins acknowledges that our biological nature will not help whatsoever in acquiring generosity. Yet he also says that those of us who overcome our natural instincts can teach our young (who are born selfish) generosity and altruism.

But if we are biologically inclined to selfishness, in order to survive, how do we learn generosity so that we can teach our young?

In the world of the scientific materialist, the contradictions abound. Dawkins, uncovered, is the very apostle of atheism sounding almost Christian.

As mentioned before, sociobiology morphs into scientific materialism when it ventures from biology to philosophy. Recently, I have realized that sociobiology has also entered the field of psychology. The objective is to fuse sociobiology with psychology.

It's called "evolutionary psychology."

On Evolutionary Psychology

I ran into this school of thought when I was documenting the origins of the philosophical concept of the "natural law." In particular, I was familiarizing myself with the writings of French philosopher Jacques

Maritain, whose book *Man and the State* is like a political Bible to many European and Latin American "Christian Democrats."

As stated before, the idea that there is an ethical "natural law," meaning a code of conduct that humans can (and must) use as a guide to control our "pleasure-seeking" passions, was firmly established by the Middle Ages. It is also well established in the law – particularly in international codes that catalog the fundamental rights of humankind.

It is considered "settled science" in political philosophy, political science, criminology and law.

The arena of criminal law is particularly clear on the concept. A criminal defense based on insanity requires that the mental health experts prove that the defendant is unable to distinguish "right from wrong." That test implies a consensus as to what is morally right and wrong. And that consensus is the natural law.

I should add that a similar distinction exists, in the criminal law, between crimes that are "malum in se" (intrinsically wrong) and "malum prohibitum" (wrong only because the law says so). The criminologist, like the other social scientists, readily admits that there is a natural law.

Once again, the only holdouts are the sociobiologists and their new cousins – the "evolutionary psychologists." Here's how Jerry Coyne, in an article in *The New Republic* (April 2000) describes the principal tenet of this new science:

> *Most of its adherents are convinced that virtually every human action or feeling, including depression, homosexuality, religion, and consciousness, was put directly into our brains by natural selection. In this view, evolution becomes the key - the only key - that can unlock our humanity.*

The title of Coyne's article (*Of Vice and Men: The Fairy Tales of Evolutionary Psychology*) is itself an indictment of evolutionary psychology. Here's a quote that contains what is perhaps his most eloquent, critical and definitive quote:

> *Unfortunately, evolutionary psychologists routinely confuse theory and speculation. Unlike bones, behavior does*

not fossilize, and understanding its evolution involves concocting stories that sound plausible but are hard to test... If evolutionary biology is a soft science, then evolutionary psychology is its flabby underbelly.

Evolutionary psychology notwithstanding, the reality is that modern society, particularly but not exclusively in the West, has blended together the best elements of religion and science, arriving at a concept of humankind that is guided by empathy as much as by selfishness.

The great ideas of the Scholastics, in the so-called "Middle Ages," flowed into the cultural eras known as the "Renaissance" and the "Enlightenment," when our species found a happy fusion of reason, science and spirituality.

The Modern Synthesis of Science and Religion

The intellectual synthesis crafted by the mind of Aquinas and his "Scholastics" ultimately led to the flowering of human reason which we call the Enlightenment. Here's how Ben Shapiro describes the felicitous way in which science and religion worked together to *enlighten* Western society as to our nature:

> *Contrary to the propaganda of a postmodern atheist movement, nearly every great scientist up until the age of Darwin was religious. The Scholastic movement produced the earliest roots of the scientific method, all the way up through the discovery by Nicholas Copernicus (1473-1543) of a heliocentric solar system.*

Science and technology picked up momentum in the last four centuries. Many mysteries were solved. Slowly but surely, all superstitions were discarded, and all insights that came from empirical science, or from within the human psyche (in many cases inspired by biblical texts), were exposed to the light of reason. As St. Augustine had stated roughly 1,000 years before, if the Bible seemed to contradict reason, it must mean that we were misunderstanding the Bible.

Science and faith were easily fused together; it was said that science concentrated on the "how" of things and religion on the "why" of things. Humanity thrived in an atmosphere of holistic discernment, represented by thinkers like Dante, Erasmus and Thomas More in the domain of humanities, Copernicus and Galileo in astronomy, and DaVinci and Michelangelo in the plastic arts.

The Enlightenment is thought to have reached its peak with the French philosophers. They added many insights to the social sciences, particularly political science. It culminated in the writings of Emile Voltaire (1694-1778); and he phrased the consensus of that era rather clearly, in relation to the importance of a supreme, intelligent being:

> *It is perfectly evident to my mind that there exists a necessary, eternal, supreme and intelligent being. This is no matter of faith, but of reason.*

Voltaire's view was shared by most of the thinkers and writers that guided the views of Western societies into the modern era. Again, their view of the spiritual world was not so much the biblical God, complete with white beard, three-prong-personality and the ability to strike you with a bolt of lightning.

It was more what is called "Deism."

Deism is merely the common denominator of the world's main religions. It is faith without rituals, or priests or dogmas. It is very similar to what later came to be called "natural theology." (We will discuss that analytical framework a little later.)

By the time of Voltaire and the Enlightenment, humanity had congealed around the notion that there is a benevolent God. That notion was accompanied by an understanding that no human is expendable and that the most vulnerable must be helped by the most powerful.

A Divergence in the European and American Models

The idea that all humans are entitled to freedom, equality and a measure of fraternity fueled liberation movements. Interestingly, in the case of the French Revolution, the belief in God was discarded.

This was not the case at all for the American Revolution, as we shall see, which was firmly premised on a belief in inalienable, God-given human rights.

England was following the lead of the French "philosophes" until well into the 20th century. Then an interesting thing happened.

In England, the holistic fusion of religion and science, which was in decadence in France and other parts of Europe, enjoyed a 20th century renaissance buttressed by a duo of natural theologians named G.K. Chesterton and C. S. Lewis. They were accompanied by another great writer, who couched his Christian views in eloquent fiction.

His name was J.R.R. Tolkien.

Chesterton and Lewis combined fiction with non-fiction to enlighten their readers as to the importance of godliness to human happiness and societal harmony. Tolkien was all fiction; and his *Lord of the Rings* trilogy was a marvel of insights as to human weakness, represented by a magical ring (emblematic of the allure of power) and human strength of will, represented by a lowly "hobbit" whose figure is more or less equivalent to the working men and women who make society great, because of their humility and instinctive ethic.

Following the British trio of thinkers, and picking up the flag of Scholasticism, which had been dropped by the secular French philosophers, a new French savant appeared on the scene. His name was Jacques Maritain; and he understood not only the relevance, but the vibrancy, of American exceptionalism.

Maritain and The UN Charter of Human Rights

As a philosopher, Jacques Maritain had special insights; unfortunately, some of them went way above our heads.

His distinction of two kinds of natural law (the "gnoseological" and the "ontological") seemed that way to me. But then I thought about it some more and it made sense. The natural law flows from both custom (gnoseological) and from logic (ontological).

Thankfully, other authors have explained that distinction in more colloquial terms. One such author is C.M.A. McCauliff, who is a professor at Seton Hall University School of Law. In a brilliant law review article, she refers to the two ways of discerning the natural law as "cognition" and "consensus." (The entire article is available in the web at 54 Villanova Law Review 435.)

Here's a quote from her article, which describes how certain self-evident truths have now found rational support in the various social sciences:

> *Maritain…described the origins of our ethical intuitions without denying that they are later given rational expression, much as the new scientists of human nature are demonstrating today.*

One of the sciences to which she refers (neuroscience), is discussed at length in an earlier chapter. Here I want to sketch briefly how the integral humanism of Maritain – itself an amalgamation of modern political philosophy with Aquinas's Scholasticism – provided a roadmap for the United Nation's Declaration of Human Rights, whose initial draft was greatly influenced by Maritain.

Again quoting McCauliff:

> *In 'The Rights of Man and the Natural Law,' Maritain wrote in 1942 that while the French Declaration of Human Rights in 1789 afforded some rights, it was based on an incorrect rationalist perspective that excluded God and looked toward to the state as the source of liberty. On the other hand, the American Declaration of Independence more closely adhered to the original…character of human rights.*

Here's the syllogism, from the mouth of Maritain himself:

> *Because we are enmeshed in the universal order, in the laws and regulations of the cosmos and of the immense family of created natures (and finally in the order of creative wisdom)… because we have at the same time the privilege of sharing in spiritual nature we possess rights vis-à-vis other men and all the assemblage of creatures.*

Jacques Maritain effectively grafted the holistic concept of human nature onto democratic ideals. His *magnum opus* on Judeo-Christian based democracy is titled *Man and the State*. It is worth reading for those who like politics explained in philosophical terms.

Furthermore, Maritain was enamored of the American experiment. He would have had no problem defending the concept of "American Exceptionalism."

The same was true for another Frenchman, who wrote in the prior century. His name was Alexis DeTocqueville.

Alexis DeTocqueville on American Democracy

For those who prefer politics explained in more clear prose than Maritain's, the best political scientist to read is Alexis DeTocqueville. His popular analysis of civic life in the United States is titled *Democracy in America*. In it, DeTocqueville narrates the miracle of a frontier nation, whose European settlers had fled monarchical systems and religious intolerance to found a nation based on individual human rights and the principle of one-man-one vote.

Here's his description of what these hardy people accomplished:

> *On the continent of Europe at the beginning of the seventeenth century, absolute monarchy stood triumphant on the ruins of the oligarchic and feudal liberty of the Middle Ages. At the heart of this splendor and literary excellence, the idea of rights had perhaps never been more entirely neglected. Never had nations enjoyed less political activity. Never had the idea of true liberty less preoccupied people's minds. It was at the same time that these same principles, unknown or neglected by European nations, were being proclaimed in the deserts of the New World to become a future symbol of a great nation. The boldest theories of the human mind were distilled into practice in the apparently humble society which probably no statesman would have even bothered to consider.*

In the same vein – three quarters of a century earlier in *Commmon Sense* - Thomas Payne had written of the American Revolution:

The sun never shined on a cause of greater worth. Tis not the affair of a city, a country, a province or a kingdom, but of a continent – of at least one-eighth part of the habitable globe. Tis not the concern of a day, a year or an age. Posterity are virtually involved in the contest, and will be more or less affected, even to the end of time, by the proceedings now.

As predicted by Thomas Payne in the eighteenth century, described by DeTocqueville in the nineteenth, and observed by Maritain in the twentieth, the intellectual fusion of democratic ideals with Judeo-Christian values gave rise to a democratic colossus, called the United States of America.

Moreover, the success of the United States fueled a host of Christian Democratic movements throughout the world. Those movements helped to lift Germany out of the evil Nazi quagmire and propel it into a modern, industrialized democracy. They helped Costa Rica become the most stable nation in Central America and Chile overcome the rightist dictatorship of Pinochet, transforming it into the most advanced of all societies in the entire continent of South America.

The same ideals are embedded in the United Nations Charter of Human Rights; they embody both commutative and distributive (social) justice. They are the bulwark of civilized societies. As we have shown, they are harmonious fusions of a proper, scientific understanding of our species' troika of body, mind and spirit. No serious historian can doubt that correlation.

Furthermore, the same philosophical compatibility that exists between Judeo-Christian tenets and the most egalitarian and prosperous political systems is also found in the domain of the so-called "hard sciences." Empirical scientists in the most advanced fields find themselves enthralled by the insights of religion, as they shed light on the latest findings of science.

Two of those, with whom I end this chapter, are Deepak Chopra and Buckminster Fuller. They are two of the brightest minds of modernity, and each has regaled us with insights that combine physics with metaphysics.

Chopra is by training a medical doctor. Fuller was an engineer and an architect. Both are eminent thinkers.

Deepak Chopra

Dr. Chopra has written more than sixty books. Many if not most have a common theme: the unification of science and spirituality.

Here's how he explained his own evolving understanding of the divine:

> *I'm coming to the conclusion that there is a difference between perception reality and fundamental reality....The latter is invisible and is a divine intelligence. Our understanding of it is based on our level of consciousness. Historically, people have had glimpses of this larger invisible, intelligent consciousness, the mystery that we call God, but they couched it in the language of their time and culture.*

Depak Chopra is not for everyone. His approach to faith is quite mystical. It is evocative of the great theologian-anthropologist, Pierre Teilhard de Chardin.

But he is emblematic of scientists in all fields who objectively and diligently seek to understand the synergistic way in which organic life functions. Like the great biologist, Pierre Paul Grasse, and like many, if not most astrophysicists and the great majority of mathematicians (what we can call "quantitative scientists"), when they observe the complexity and fine-tuning of the universe, they are propelled to reach for metaphysical explanations.

The same is true of physicians, who constantly "reverse-engineer" the human body in an effort to understand illnesses and cure them. And it's also true – and perhaps more poignantly – for engineers, who marvel at the ingenious crafting of our planet.

Moving from Pierre Paul Grasse, in biology/zoology, to engineering and architecture, we encounter one of the most respected thinkers of the last century: Buckminster Fuller. I was privileged to see and listen to this modern polymath, who is credited with coining the word "synergistic" to describe phenomena in which matter, mind and spirit interact so that the whole is greater than the sum of the material parts.

At Harvard University in the early seventies, this brilliant architect, engineer, and inventor regaled us with his insights. I found out later something he didn't mention at Harvard – that he had crafted a scienfitic version of the Lord's Prayer....

Buckminster Fuller on the Lord's Prayer

Fuller's version of the "Our Father" begins by stating that "to be satisfactory to science, all definitions must be stated in terms of experience." In other words, science to him must be empirical.

Then he proceeds to set some parameters. "In using the word God," he explains, "I am consciously employing...four experience-engendered thoughts." The four are:

1. "All the experientially explained or explainable answers to all questions of all time." (In other words, the complete body of empirical science.)

2. "The individual's memory of many surprising moments which... engender the reasonable assumption...of a total comprehension of the integrated significance...of all experiences." (In other words, our human intuition tells us that all our experiences, from birth to death and perhaps beyond, are somehow interrelated and have coherent logic.)

3. "The generalized principles...thus far discovered and codified... and ever physically redemonstrable by scientists to be not only unfailingly operative but to be in eternal omni-interconsiderate, omni-interaccommodative...ultra-tunable micro and macro-Universe events." (In other words, what many scientists refer to as the fine-tuned universe, which begs for the conclusion that an intelligent designer created it. A good book to read on that is *The Privileged Planet* by Guillermo Gonzalez and Jay Richards .)

4. "All the mystery inherent in all human experience, which as a lifetime ratioed to eternity is individually limited to almost negligible..." (In other words, the instinctive recognition that human experiences are quantitatively minimal compared to eternity, which leads to the conclusion that we share the universe with a much larger intellect-creator.)

By now, I hope I haven't lost my reader to the arcane, multi-discipline, overly-nuanced and technical eloquence of the planet's last polymath. So I will cut to the chase and offer my reader Fuller's emotional reaction when he glances at the glory of creation and marvels at the power of its creator.

Reverting to a much less intellectual mode of speech, Fuller ends his scientific version of the Lord's Prayer more simply:

> *You, Dear God, are the totally loving intellect…wherefore we have absolute faith and trust in You, and we worship You, awe-inspiredly, all-thankfully, rejoicingly, lovingly, Amen*

And thus, in the end, Fuller resorts to more lyrical language in a valiant attempt to explain divine love. He was truly a "beautiful mind."

Another twentieth-century "beautiful mind" was John Forbes Nash. The next chapter leads off with his description of what is clearly the least quantifiable, but most important of all human traits.

As he described his struggle with schizophrenia, it was love that rescued him.

CHAPTER VIII
FAMILY: CRUCIBLE OF LOVE

My quest has taken me from the physical to the metaphysical, the delusional and back; I have made the most important discovery of my career – the most important discovery of my life. It is only in the mysterious equations of love that any logical reasons can be found. John Forbes Nash, on receiving the Nobel Prize for Economics (as depicted in the movie "A Beautiful Mind.")

Nash was one of the great minds of the twentieth century. He is credited with inventing the science of Game Theory, which is a sub-discipline of economics. Game Theory offers mathematical proof that the welfare of the group must often be weighed as much as the welfare of individuals. We cannot maximize individual welfare if we totally ignore the common good.

What is true in economics is also true psychologically and sociologically. Generosity of spirit is a trait that nurtures the human psyche. For reasons that beg for explanation from something that is not readily apparent, the human at birth is not all that inclined to share.

Love has to be taught. And it is initially learned by humans at the foot of their parents' bed. In the case of motherly love, it is learned right from conception.

The well structured family is like the edifice in which human, emotional stability dwells. The parents are like the columns in that edifice and the love between the parents are like the connecting beams that support the floors of the structure, on which the children are able to thrive.

Continuing the metaphor, we can compare the expression of sexual love that binds the parents to the cement that solidifies the floor slabs. As discussed in Chapter V, the expression of sexual love in a stable marriage is not only one of the most joyful of human pleasures, but also arguably the most fulfilling.

The reason why it's fulfilling is that its salutary effect spills over into the other dimensions of human existence. The love and affection of the parents for each other, particularly when it is in evidence for the children to observe, is emotional nourishment for them. It's like a triangular exchange that soothes and solidifies any cracks in the emotional development of the offspring.

This is not only self-evident but supported by myriad sociological studies. No psychologist or social scientist doubts that a stable family unit is conducive to stable development of the children thereof. When there is no stable family, there is mayhem.

We use the term "broken" to characterize a family that is no longer a connected unit.

On Broken Families

Every study that has analyzed this phenomenon reaches the same conclusion: a family that splits at the top results in harm to those at the bottom. For those of us involved in municipal government, the statistical reality of the modern, "broken" family is a worrisome condition, to put it mildly.

About half a century ago, a liberal, Democratic Senator named Daniel Patrick Moynihan published an exhaustive study of the effect of broken families on the African-American communities. It was not particularly well received among black leaders, who saw it as a sort of justification for the economic inequality of American blacks.

Nowadays, the divorce rate has skyrocketed among all races and ethnicities. More than half of all marriages end in divorce. And the number of children who are born and raised by one parent is enormous.

The consequences are particularly felt at the level of primary education, where the children from broken homes are statistically at risk from the moment they arrive in first grade.

My wife is a first grade teacher in one of the best public schools in the area – perhaps in the nation. One time, she was sharing with me the travails of a youngster who was having a really tough time "adjusting," which is a euphemism for doing course work. The student was having difficulties learning to read and write.

I asked her if the child was from a broken home; she answered: "Honey, every child who is struggling with school is from a broken home."

She didn't really mean to imply that there was a 100% correlation between kids having difficulties and failed marriages. There are all kinds of causes for educational dysfunction.

To begin with, there is the obvious genetic deficiencies – or what today are called "special needs." Children with genetic deficiencies have a hard-time adjusting to school, whether it's attention deficit disorder, autism, low I.Q., or other special need such as poor eyesight/blindness/deafness.

Others have developmental deficiencies and those are a function of parenting, sibling rivalries, and communal deficiencies such as might cause under-nourishing and/or under-nurturing. It is in that realm (personality development of the growing child) where the stability of the marriage has its greatest impact.

In effect, what my wife meant was that every single child who came from a broken home inevitably faces obstacles to healthy personality development and the achievement of academic and emotional advancement. Moreover, what used to be a dichotomy (nature versus nurture) is evolving towards an analysis that begs for at least one more component.

It would behoove social scientists to stop talking as if there is much that could be called "settled science" in this area; the only thing everyone agrees on is that human personality development is not purely genetic. And that leads us to consider the latest in psycho-social research, beginning with what is perhaps the most studied of human traits: cognitive skills.

The Fifty-Fifty Rule in Cognitive Development

Just about every study that has been carried out suggests that the purely hereditary component of human intelligence is at most 50% of the total equation.

In her book analyzing intelligence deficiencies among children of underprivileged families, cognitive theorist Ruby Payne (*Understanding Learning: The How, the Why, the What,* 2002) quotes Israeli cognitive psychologist, Reuven Feuerstein, when he famously pronounced that "it is possible to have a brain and not have a mind. A brain is inherited, a mind is developed."

In what sounds like a perfect quantitative arrangement, Payne puts it in its simplest terms:

> *Cognitive scientists have concluded that it's about a 50-50 arrangement. About half of who an individual becomes is developed by his/her genetic code and about half by his/her environment.*

Modern psychologists extend the 50-50 model well beyond intelligence. As Judith Rich Harris explains it, "the correlation of personality traits (as estimated by scores on personality tests and in various other ways) is only about .50 for identical twins raised in the same home." Since nature and nurture are the sum total of variables considered by most modern scientists, the other 50% is attributed to the environment (a/k/a "nurture").

Alas, it is not that simple. In their zest to eliminate factors that are not easy to measure, scientists have over-reached and made it sound like our species is like a football field, wherein fifty yards are marked for the home team and fifty yards for the opponent. Nature (biology) has the first fifty yards, and the environment (nurture) the other fifty yards. They make it seem as if the two (nature and nurture) add up to a touchdown – each and every time.

It is not quite that tidy, particularly as pertains to nurture.

Assuming for the moment that there is clear and universal proof that intelligence is correlated to heredity by a 50% factor, what says that the other 50% are determined by one's environment – or what psychologists call "nurture"? There is really no solid proof, yet the literature overwhelmingly treats it as if it were settled science.

For the sociologists and the educators, it appears as a quantitatively airtight solution: Our task as a society is to ensure that all children are born physiologically healthy. The second task, which is shared by parents and educators, is to make sure we allocate educational goodies equally. Premised on those two conditions, we can expect that the achievement curve will follow the statistical, genetic "Bell Curve" and we can feel good that we have given every child the same opportunity that nature gave him/her.

Into that tidy, quantitative framework came family therapists and just about convinced us that the "nurture fifty" depended almost entirely on the stability of the traditional family. Give me a daddy, a mommy, and a decent, well-kept home and you have the perfect formula for success.

The formula has at least one important missing variable; it is the person's unique power of decision, otherwise known as free will. But there are others. For example, there is increasing evidence that the influence of peers and mentors is as important, if not more, than the influence of biological parents.

Even so, there is no denying that a missing parent is often a hurdle to be overcome on a child's way to maturity and happiness.

Nature, Nurture, and None of the Above

As Judith Rich Harris explains it, "the most famous – and most pessimistic – study of the children of divorce is the one by clinical psychologist Judith Wallerstein...." But Harris cautions us:

> *Wallerstein found a very high rate of emotional disturbance among the children of middle-class divorced couples. Her books sold a lot of copies but as science they are useless: all the families she studies had sought counseling and all were getting divorced. There was no control group of intact or self-sufficient families with which to compare the children of her patients and no way to filter out her professional biases.*

Harris proceeds to cite a "properly controlled recent study of the children of divorce" that "gives a more optimistic picture." It came from a massive British survey of children born in a single week in 1958; the study was done in 1981, when they were 23 years old.

The British study found a very slight increase in the incidence of depression among the children of divorced parents. However, most of the studies she cites are done among middle-class, British and American parents.

The rate of divorce is much worse among minorities in the United States – particularly African-Americans; in that demographic, the number of children who are raised by a single mother is approaching 70%. A lot of studies will have to be done to determine, in today's single-parent society, how much damage is done by what is now the "new normal," i.e., the disappearance of the traditional family, composed of mom, dad and their direct offspring.

In any case, divorce is but one variable in the nurture part of the formula for raising children into maturity. Harris's main thesis is that our species is influenced only slightly by the family environment. She accepts as a given a host of studies tending to show that pure heredity "accounts for about half the of the variation among the individuals."

Well, we knew that. But what about the other half? Is it all nurture? No, said Harris, we need to look at three components: nature, nurture, plus "none of the above."

Her thesis finds support in the popular saying that the child is raised by the entire tribe. But what if the entire tribe is broken?

The Broken Village

It has been said that "it takes a village to raise a child." Maybe it would be better to say that *it takes a family plus a village.*

One thing is for sure. When both the family and the village are broken, chaos ensues.

For various reasons, including government neglect, Miami's Overtown neighborhood was such a neighborhood. At one point, more than 95% of its residents were renters, rather than owners; more than 85% of its

126

households were headed by a single mom; and almost 100% lived in poverty.

Children growing up in Overtown had no chance of success, unless their parents could somehow make it out of the "ghetto" and away from ubiquitous drug dealers, rampant crime, and schools where many, if not most, of the students were promoted from year to year without actually learning to read and write proficiently.

Redressing this state of affairs was not easy. As far as I know only the United States – among the industrialized nations – experienced this disparate existence, in which the middle class emigrated to the suburbs and the inner city was left to deteriorate into what legally is called "slum and blighted" condition.

Perhaps for that reason, the United States is the only highly developed country that has tackled endemic societal dysfunction within its own borders. At least in Miami, it is succeeding to a great extent in compensating for the broken family with a variety of child-care facilities and parent-support systems.

The Parent Substitute: *in loco parentis*

More than a decade ago, we approved, in Mami-Dade, a special tax to fund early childhood facilities. It provides well over $100 million each year; and allows us to claim that "no child is left behind" in terms of access to pre-school educational facilities – which also provide a fair amount of nutrition.

Our specially funded agency is called the Children's Trust. I served on it for two years, and managed to advance two initiatives which I consider absolutely vital to the formation of youngsters.

One program sponsored by the Children's Trust seemed a bit suspect to me, when I found out about it. It was called "parenting support." At first blush, it seemed superfluous, if not downright presumptuous, to think that the state can hire people to be substitute parents. Then I realized that taking kids away from their natural parents is a more drastic (and often counterproductive) measure when they are being neglected than supporting the parents themselves – preferably in their existing home environment.

The Miami-Dade Children's Trust is a wholesale, societally funded effort to prepare children for school and to enhance the educational experience. It provides pre-school and after-school instruction, nutrition and nurturing to both children and adolescents. It is a lifesaver for enormous numbers of minors whose parents are not there before and after school.

Buttressed by early childhood initiatives like the above, and combined with the entire public school system, the children in our county do well until adolescence. Up to that point, the village complements the family well.

Then all hell breaks loose.

Let's delve a little more into the first phase of childhood development, which, as a general rule, is relatively successful in the United States.

From Infancy to Adolescence

In the introduction to the Judith Rich Harris book previously quoted, Harvard's Steven Pinker refers to the "dogma" that a baby's "attachment" to its mother "sets the pattern for its later commerce with the world." Pinker is harsh in his judgment of what he calls a "tired notion bequeathed to us by Freud." He expands as follows:

> *Relationships with parents, with siblings, with peers and with strangers could not be more different, and the trillion-synapse human brain is hardly short of the computational power it would take to keep each one in a separate mental account.*

In other words, the process of programming the human brain, from birth to adolescence, is much more complicated than anything that Sigmund Freud could possibly have formulated, supported by the rudimentary science of his time. Perhaps we can say (using the football analogy) that *the mother is the most valuable player in the race for that last fifty yards* of the playing field that marks the child's development; but she is not the be-all and end-all.

Still, when the entire childhood-to-adulthood journey is analyzed, there is no denying that mothers have influenced, genetically and otherwise, both the "hardwire" and the emotional programming ("software") necessary for the child to deal with the environment outside the womb.

In this regard, it is worth noting that the female in the human species seems exceptionally well equipped herself on the traits necessary to raise a child that is healthy physically and psychologically.

For example, mothers instinctively know enough to breast-feed a child; through most of history, the signal for feeding was the baby's cry. Since the mother had no way of measuring how much milk the baby was absorbing, it was feed-on-demand. (Nowadays, in many advanced societies, there are artificial means of supplying sustenance, including pumps and bottles that can store the mother's milk, containers to measure volume and clocks to measure the time span between feedings.)

There is also a gadget called a "pacifier," which tricks the child into thinking he/she is still sucking nourishment and induces sleep – or at least quiet – so that parents can get a break.

As much as any other human being, except possibly a professional in the child-care industry, I have substantial experience raising kids.

As the saying goes, there is no "manual" that comes with a marriage license or for couples who otherwise begin the experience of fatherhood or motherhood. The lack of a working manual becomes a real quandary when children reach the magical age of adolescence.

My Personal Book on Adolescents.

A lot of what I know about how to raise kids is purely anecdotal. Most of it has come from experience in taking over as parent for the last five siblings in my family. Then came the experience of raising four of my own kids and babysitting my eleven grandchildren. In between those experiences were others helping to nurture, advise and coach some fifty-two nephews and nieces.

The two main laws of nurturing children that I have learned can be stated as follows. Rule No. 1: Instill discipline into the infant child, gently but firmly. Rule No. 2: Gradually allow the child to make his/her own decisions, such that by the time of adolescence, the discipline is self-imposed.

Many of my close relatives (that I will not dare to identify) fail to abide by Rule No. 1. They give the child way too many choices. Instead of saying to a three-year-old: "Honey, it's time to go home after a great day at the beach," they will say: "Do you want to go home, now?" Or instead of saying, "Honey, I am baking you some delicious chicken nuggets," they will say: "What would you like for dinner?"

On the second example above, I often cringe when the question is asked – knowing that there is nothing but chicken nuggets that can be prepared instantly; anything else would have to be ordered from a nearby restaurant.

Besides a lot of experience with young children and adolescents, I have a fair amount of theoretical and practical knowledge of a few personality disorders. I have dealt with separation anxiety depression. I have advised and read extensively about autism and the related Asperger's Disease. I have been a part of a 14-subject, in-the-flesh experience in maternal narcissism. (I refer to my various experiences with my mom and our fourteen siblings.)

Buttressed by theory and experience I can offer some practical advice to parents and educators regarding what used to be called the "formation" of the child's character. Kids have to be taught not to be self-centered and self-absorbed. Kids have to learn, sometimes the hard way, that they are not the center of the universe.

Moreover, kids need to have coaching on how to exercise their own free will. And it helps enormously if they have a sense that they are not only your children, but children of a higher being. This becomes more and more important as they reach the age when they have freedom to choose.

Here's why. Again, let me stress that my observations are purely anecdotal.

On Children and God

I have found no documentary support, no clinical studies, and nothing published on the insights I offer here – although I suspect they exist.

It is clear to me that kids need a figure higher than their parents to turn to, when things go wrong, and to thank, when things go right. Let's take the latter scenario first: when things go right.

You see the effect when families pray together to thank God for the meal in which they are about to partake. The classic prayers says something along the lines of "bless us O Lord, for these thy gifts…"

The children gathered around a table seem to reflect on this mythical scenario, in which they are encouraged to thank an invisible benefactor who is more powerful than their parents. This idea is probably not as important when things are going well. Even so, thanking the creator of all things for "gifts" of the earth is probably helpful psychologically.

For one thing, it gives credit to the presumed parent of all the earth's children and thus, implicitly, suggests that all children are equally entitled to basic sustenance. In our family we also pray for those who do not have enough to eat – *thus reminding the little ones that it is our obligation to help make sure all children have the necessities of life.*

Now let's look at the other half of the coin: when things are not going so well.

Praying to an almighty, all-wise and all-loving uber-parental figure has great benefits, psychologically speaking, when things are not going well. This has to be particularly impacting for kids who feel mistreated or misjudged by their earthly parents. In those circumstances, *God is like an appellate judge*: The child internalizes the hurt, but does not have to suppress it entirely. There is a voice of reason that will listen to the spurned child – that will always treat him or her fairly. It is reassuring to the ego that there is something akin to an appellate court that metes out justice more equitably than one's own parents….

These observations are purely anecdotal. Another one that is anecdotal, but clearly on point, is the balance between discipline and affection in raising children.

Discipline Versus Affection

I have discussed this issue with many of my brothers and sisters – as well as a psychologist or two. My conclusion is shared by everyone with whom I have discussed it.

It is almost self-evident.

And yet it is not universally put into practice; far from it. Many young couples struggle to find the right balance between discipline and affection.

It bears repeating: Discipline is important; a child who does not learn the value of it will have a hard time succeeding once he/she is alone in the cruel world, without parents, teachers or coaches to guide him/her.

Having said that, it is important to note that affection is more important to the child than discipline. *If one is going to err on the side of too much affection without discipline versus too much discipline without affection, it is clear that the worse option is too much discipline.*

In the final analysis, a child needs love more than anything else. All the experts agree on that.

Often it is needed to repair inborn or developmental voids in self-esteem. Psychologists have catalogued a host of mental health issues that plague the child who lacks in self-esteem. One of those is called "separation anxiety depression;" it is a relatively new diagnosis for the youngster who seems to be unable to feel happy at home, but also unable to function away from home. He runs away from home to escape the depression and finds out that the depression is worse when away from his parents.

This book is not the place to delve into that condition, except to note that it manifests itself early in life (as early as seven years old) and that parental love, and its outward expression that we call affection, is the single most important cure.

Other voids in self-esteem manifest themselves later in life. One that seems to affect girls more than boys, when they first move away from home (typically when they go to college in another city), is worth mentioning, because it is so widespread.

Depression in the College-Bound Youngster

In tight-knit communities such as Miami's Cuban-American community, now numbering about three-quarters of a million people, a geographic break from family is quite traumatic to the youngster.

Let me illustrate.

Cuban-American parents in Miami tend to encourage their kids to stay home for college. That is partially an economic decision, but more often it is simply the desire of parents to have their kids at home, even when they reach the societally accepted age of adulthood, which is 18.

That age, particularly among girls, is precisely the statistical onset of a dysfunction that affects one out of every five young women in America.

The statistics are startling for Americans generally: as many as 17 million Americans suffer at least one episode of acute depression that impairs their normal functions. That's about 5% of the population of the United States.

Among young women, based on 2017 statistics, the figure is much higher, at 20%.

I am not competent to analyze the causes of that startling statistic. Clearly, many of the cases of depression are "situational," meaning that they are prompted by an external occurrence, as opposed to something inherent in the person affected.

I saw the phenomenon first hand when two of my daughters went away to college. I have three daughters, and only one did her undergraduate studies while living at home. She never complained about being depressed. The other two went away to college, and lived with other co-eds in dormitories or apartments close to the college campus. They both showed symptoms of depression – which thankfully were short-lived. All three of my daughters were stable emotionally and successful academically and socially. Anecdotally at least, it begs for a correlation that connects the depressive episode with leaving the family nest.

I am not aware of any studies that can establish the correlation with any degree of scientific confidence. But again, anecdotally, it seems to be a function of leaving the home for college for the first time – particularly in the case of young women, given that the rate is about three times higher for them than for young men in that age group (18-25).

I have to assume that the disparity between females and males is also related to the different ways in which the body, mind and spirit adapt to the onset of sexual attraction. Psychologists, educators and parents would do well to combine forces to understand this phenomenon, which affects all youngsters – particularly young women.

In an earlier chapter, I discussed the related dysfunction of clinical depression, which affects fully matured adults. There is no doubt that the human species is unique in this rather widespread trait, which has roots that science is still trying to understand and treat.

Undoubtedly, the person who is suffering from severe mental health dysfunction has more difficulty socializing, identifying with, and feeling altruistic towards others. As in the famous book by Thomas Anthony

Harris (*I'm OK; You're OK*), it all starts with healing one's inner psyche; ideally that is when one is ready to love others.

But there is a paradox here. Love is not what sociologists and game theorists call an "on-off" or "zero-sum" variable. Love does not diminish with use; to the contrary, it increases with use.

The psychologically needy person also *needs to be needed*. And it doesn't really end there. All humans need to be needed. We search in vain for our twin brother or sister, our "alter-ego" with whom we can share thoughts, aspirations, longings.

Most of us, most of the time, yearn for the love of equals.

That brings us to Plato.

Brotherly (Platonic) Love

Philosophers going back to Plato recognize that there are many kinds of love. The most popular categorization subdivides love into *eros*, which is equivalent to sexual or romantic love, *agape*, which is filial love, and *caritas*, which is the most altruistic, most selfless, and least comprehensible to the scientific materialists.

Plato gave us a special insights into the love that binds friends or brothers and sisters. In Plato's view, the most perfect love can only exist between two equals; in that sense, the bond that exists between me and my siblings or my close friends is the more exquisite the more we share views, age, beliefs and pursuits.

It is easy for me to have a special bond with brothers and sisters who have college degrees, believe in a higher being along the lines of the Judeo-Christian God, and are convinced that all members of our species, regardless of race, color or creed, have the same basic rights.

It helps, in my case, if my friend or brother is also somewhat tuned into sports and politics. Sports, for me, are a form of entertainment. Politics, for me, is the stuff of life. And that is not just a figure of speech. St. Augustine, who is arguably one of history's three greatest philosophers (along with Aristotle and Aquinas) once said that politicians have the highest calling – even higher than priests, ministers and rabbis.

Leadership of the entire society entails a combination of human traits with a good dose of economic resources and a lot of luck. The personality of well-known leaders is easier to analyze for the simple reason that it is the one about which we know the most.

In the next chapter, I will talk about human personalities. Psychology and psychiatry deal mostly with pathological personalities. With the exception of sinister leaders like Hitler, Stalin, Pol Pot, the North Korean Un dynasty and the Cuban Castro brothers, the personalities of leaders is a study in mostly beneficial traits.

So is the average human, exemplified by the typical relative or friend. Most humans are good people; most leaders represent the best of our species. But good people come in a lot of sizes and colors. Some are interesting and lively; some quiet and studious; some do their best to combine all positive traits.

In the next chapter, we attempt to shed light on the whole panoply of human personalities.

The popular literature is replete with references to "alpha" males and females. Also mentioned is the dichotomy between people who are introverted and those who are extroverted.

Those are important, binomial variables. Yet the reality is much more diverse – more akin to a matrix than a binomial equation. In terms of empirical science, the closest analogy I can think of is to "chaoplexity," which is simply the combination of chaos and complexity.

It behooves us to tackle the study of personality, using all the tools of modern science.

CHAPTER IX
OF HUMAN PERSONALITY

Personality is something that we all have. It's that unique organization of one's strengths and capabilities, psychological defenses, developmental adaptations, styles of coping as well as genetic predispositions. Russ Federman, "Psychology Today," October 27, 2013.

*T*he above quote, by a respected psychologist, comes from a current paper in what is perhaps the foremost professional publication in the entire field of psychology. It is called *Psychology Today*

Let's break it down a bit. The author suggests that human personality is "uniquely organized." I think, actually, that he meant to say uniquely "combined" rather than uniquely "organized." The reason I say that is that "organizing" implies that a rational actor put the entire construct together.

Which is probably true, as we shall see in various parts of this book. So is the "unique" qualifier. Every study of identical twins, raised by the same parents in the same environment, shows that they have significantly different personalities.

Even conjoined twins develop distinct personalities.

For conventional empirical scientists of the current era, this simple fact is puzzling: Each human is unique. For us it is self-evident.

As I previously mentioned, there is a Spanish saying that goes: "Cada persona es un mundo." Roughly translated, it means that each person is unique. Each person must be analyzed by a near-infinite number of traits – some genetic (nature), some developmental (nurture) and some volitional (responding to the will) – that defines the uniqueness of each.

Moreover, the personality of a human is something that changes over time. So that begs the question of how static is personality at the beginning of the process and how much can it be adapted by the subject or his surroundings.

Modern science suggests that the process is actually quite dynamic. A lot of what Federman tells us is a "unique organization" is done by the very subject herself.

And contrary to the conventional wisdom, it is quite the case that people change.

William James: Multiple Personalities

One of the biggest mistakes that parents and educators make is to categorize the personality of children at an early age. Personalities, like occupational vocations, are the result of many factors, of which genetics and parenting are only two.

The social milieu in which we humans function, the books we read and movies we watch, the example of people we admire – all of those, woven by our own choices, complete the panoply of qualities that shape our personality.

And it changes over time, based in great part on the people with whom we spend time. In childhood and adolescence, that is mostly the classmates, teachers and coaches that surround kids during school hours.

Once we are adults, particularly after working hours, we often get to choose our peers. At this point in our lives, as described by William James, we develop "as many social selves as there are individuals who recognize [us] and who carry an image of [us] in their mind."

Here's how James describes the process by which a person develops the various different sides of himself (or herself) that cater to "distinct groups of persons about whose opinion he [she] cares." He illustrates with various examples, as follows:

Many a youth who is demure enough before his parents and teachers, swears and swaggers like a pirate among his 'tough' young friends. We do not show ourselves to our children as to our club-companions, to our customers as to the laborers we employ, to our own masters and employers as to our intimate friends. From this there results....a perfectly harmonious division of labor, as where one tender to his children is stern to the soldiers or prisoners under his command.

This reality, which begs for an explanation that includes the use of our free will, is as evident as any law of physics or any math equation. We shape and add to our aptitudes, discipline our passions, and emulate those who inspire us.

Many of us do it by what is called prayer.

Prayer and Personality

What James says about how our species caters to various groups whose opinion we care for is also applicable to how many of our species adjust their behavior to fit what God might expect of them.

The great civil rights leader, Dr. Martin Luther King, once said that when he was an unknown minister, he tried to spend an hour a day on prayer and reflection. Later, when he became the national leader in that struggle, he thought he would have to cut short the time spent in prayer – only to realize that the demands of his leadership role actually *required two hours a day of prayer, rather than one.*

In my current legislative role, when I must sit through endless county commission meetings in which there is a lot of grandstanding by both legislators and citizens, I often resort to prayer.

Like the poet, I must "count the ways" in which prayer helps me adjust and motivate what might appear to others as my inherent personality, but is to a great extent an acquired one.

Turning to some higher being for inspiration helps me enormously to *put a happy face on what would otherwise be a frown.* And I mean this literally.

Having read that wearing a smile requires the use of less muscles than wearing a frown, I often take heed. I do this even when alone: I literally adjust my expression when thinking about others whom I might dislike, or whose conduct and views make me angry.

It does wonders. All of a sudden, the same person whose views or conduct is aggravating flashes in my mind as a person to whom I am inclined to show compassion, or even love.

The effort to smile at and interact with people is a constant feature in my life. Political leaders who don't smile are lost.

Leaders also have to be optimistic in the face of adversity. I call that trait "faith-based perseverance." It is not just the "right stuff." It is the "only stuff."

Many biographies of leaders emphasize this trait in the great leaders in history. Abraham Lincoln would infuse his cabinet with the thought that a victory in battle was the prerequisite to emancipating the slaves, because it would impress the citizenry with the idea that God was on the side of the Union.

Leadership comes from the recognition that the search for truth and justice must be buttressed by the belief that the citizenry can rise above petty concerns, including the internal misgivings of the leader himself.

Like the monk or the missionary, the leader dances to a different, more mystical tune. In that, he/she exemplifies a trait that distinguishes our species from the animals in a special way: the leader marches to the beat of a different drum.

There is no place for herd mentality.

Note how we draw the distinction in common parlance. We laugh at those who "follow like sheep" or "go over the precipice like lemmings." We deride those who act "like Pavlov's dog" or allow themselves to wear the kinds of "blinders" used to keep horses in line.

Perhaps more than any other human trait, this is the one rooted in the concept of free will. The decisions of a true leader are often contradictory, and often result in the unpredictable juxtaposition of the leader's instinct to survive and his/her desire to push the envelope of probabilities towards what is often a seemingly impossible dream.

Let's delve a little bit into the personality of a leader.

The Personality of a Leader

I am not even remotely competent to analyze, let alone diagnose in any scientifically rigorous way, the personality of dictators or tyrants. In the case of Adolf Hitler, a multitude of writers, thinkers and psychologists have analyzed his macabre personality traits.

One whose book I have read is Hanna Arendt, and she tackled, in *The Banality of Evil,* the tricky question of whether someone with normal mental health can commit mass murder. That question is analogous to the theological discussion of whether any human being deserves to suffer in some sort of eternal hell for his/her actions during life on earth.

The analysis of pure evil, or the related question of eternal damnation, is beyond my purpose. I prefer to analyze the other multitude of human personalities which have mostly good traits mixed in with some measure of narcissism.

In the case of leaders, I base my observations mostly on biographies, of which I have read more than a hundred.

I am also uniquely positioned to analyze the personality of one municipal leader: myself. In doing so, I will delve into one of the three traits that distinguish humans from animals: the one that I have previously called "faith-based persistence."

On Faith-Based Persistence

I wish I had a dollar for every decision I have made where the odds were against me. The same would apply to George Washington, Simon Bolivar, Joanne of Arc or Martin Luther King. Those are leaders whose biographies are well documented.

Similarly for Winston Churchill. I have read and seen enough of Sir Winston to have a fairly educated opinion of what made him tick. Perhaps what attracts me to Churchill is that we have similar personalities.

Various authors have argued, based on the rather extensive writings, public appearances and pronouncements by Churchill, that he was innately shy, but forced himself to be what today we call a "people person." He was a natural "nerd" who forced himself to be what psychologists call "an intuitive extrovert."

I am not a born politician. Like Churchill, I am a made politician.

Of all the leaders in history, the one I would have loved to meet is Winston Churchill. I identify with him more than any other. (I suppose a close second would be Sir Thomas More. And a close third might be former, famed New York Mayor Fiorello LaGuardia.)

Thomas More is the "Man for All Seasons" that Robert Bolt wrote about in the academy-award movie that has been seen for decades. More gave us the book "Utopia," which is now an adjective that we use to describe the ideal society. (He was not only a statesman, but was, along with Erasmus, the top intellectual in the Europe of his time.)

LaGuardia is perhaps the most famous American mayor of all times. He guided New York in wartime and peacetime like no other. A story is told about him that defines his zeal for social justice.

And so I digress to tell one LaGuardia story.

At one point in his career, LaGuardia was a municipal judge, who presided in a court that dealt with minor crimes. One day, a man was brought in with a charge of stealing meat from a grocery store. LaGuardia asked him how he would plead and the man said "guilty with an explanation."

LaGuardia asked him to explain. The defendant said that he had been unemployed for six months, and had to steal to feed his family.

LaGuardia sentenced him to the mandatory $10 fine, then took out a $10 bill from his own wallet and paid the fine to the bailiff. Whereupon he said: "Now, I find guilty and sentence every person in this courtroom for living in a city where a man cannot get a job with which to feed his family. Mr. Bailiff, collect twenty-five cents from each person in the courtroom."

All told, the bailiff collected $47.50, which LaGuardia then handed to the defendant and told him to go buy food for his family.

Stories like this inspire us to be better human beings – and in particular, better leaders. In the case of Churchill, I admire how he changed his personality so that he could lead a small island nation through two world wars, and write about that nation's gloried thousand-year history as the beacon of human rights that began with the Magna Carta in 1215.

Like Churchill, I am an intuitive, nerdy person who prefers to be alone, or in a small group, rather than the madding crowd. I prefer to read and write than to party or campaign for public office. As I write these lines, I am close to the end of my fifteenth campaign in forty years, as follows:

1979, 1981, 1983, 1985, 1987, 1989, 1996, 1997, 2001, 2004, 2006, 2011, 2012, 2016 and now, in 2020.

Those fifteen campaigns don't count what is called a "run-off," which is a whole new, much shorter campaign between the top two contenders. I had run-offs in 1981, 1983, 1985, 1987, and 2004. So that makes 20 electoral contests in total.

No one can argue that I lack persistence.

This trait must be explored a little. It is not just the thirst for public acclaim, or power; in my case, it is certainly not the thirst for financial gain – as I served as mayor of Miami for $5,000/year and as county commissioner for $6,000 per year.

It is a combination of all those traits, plus a strong dose of what I call "faith-based persistence." Once again, I turn to Churchill for a definition of persistence, which he called "the ability to move from failure to failure without losing enthusiasm."

By all accounts, at least in my case, persistence is a trait that comes from equal parts nature, nurture and inspiration. I have already dealt extensively with nature and nurture.

Inspiration is something entirely derived from outside our physical beings. It is not drawn from physical sinews or brain capacity as much as from external and internal inspiration, neither of which is rooted in a particular biological organism. It typically comes from those around us, whom we yearn to imitate. It also comes from those we read about and see in movies and plays.

And it often comes from a higher power.

Let's break those down with a little self-examination.

The Inspiration We Draw from Our Environment

Undoubtedly, for me the initial thrust of leadership inspiration came from family members. My dad was for me the initial, primary factor: He was the classic "over-achiever" who was never satisfied with knowing anything halfway or doing anything halfway.

The best example was in sports. We knew, from pictures in yearbooks and stories told by his contemporaries (occasionally embellished by my mother) that dad was a superlative athlete. He was fast and very coordinated, which allowed him to excel in track and basketball.

Somehow, by the time he attended Villanova University in Philadelphia, when he was barely 17 years old, he managed to make the varsity tennis team in one of the best sports programs in the nation. How he went from being a stellar runner and basketball player in a Cuban high-school (which didn't even have a tennis program) to playing varsity tennis at a major American university is beyond me.

There was a lot of natural talent, but the rest was "ganas," as in the famous true-to-life movie, "Stand and Deliver," which tells the story of Jaime Escalante. In the movie, Edward James Olmos plays the role of Escalante, who taught poor, Hispanic kids in East Los Angeles to excel in advanced math, when no one else thought they could perform well in any kind of math.

"Ganas" is like ambition, but perhaps more from the gut than from the mind or the heart. In my father's case the "ganas" was combined with ambition to succeed and with the thrill of competing.

The man loved to compete.

When my father reached the professional phase of his life, he yearned to be the best agronomical engineer in Cuba, which at the time was the world's biggest producer of sugar.

Later, in exile from a communist regime, he yearned to be the best at designing nuclear bomb shelters (which he did under contract with the British government) and in designing nuclear power plants (which he did while working for the world's top engineering firm in the field, named Bechtel).

A lot of his success was motivated by something that comes from outside the person – perhaps even outside the tribe. In the military they call it performing "above and beyond the call of duty. " In sports, it is often called "giving it 110%"; in other words, the ability that humans have to excel beyond any normal, measurable physical constraints.

Admittedly, and assuming my dad was like me, there is an important element of self-indulgence in the motivation to excel. My dad and I both were born with traits that allowed us to excel in sports and academics, and the urge to shine above our peers undoubtedly had elements of vanity and personal ambition.

And yet I could see in him, and feel in myself, strong external (and mostly altruistic) forces at play. Among those were the example of great leaders, including political and religious ones.

How Leaders Are Inspired

My dad particularly admired Germany's post-war prime minister, Konrad Adenauer, whom he saw as a committed Christian Democrat. He admired, and followed the teachings of, popes like Leo XIII and John XXIII who instilled into European and American culture the notion of "distributive justice," which is the key idea that takes us from pure self-interested decision-making to a compelling promotion of the common good.

As a young man, I was quite busy with academic pursuits, which were soon mixed, during my adolescent years, with the survival struggles of the exile condition.

At the age of 20 or so, I started devouring political biographies. The ones that stood out were written by Manchester, and they described political icons like Churchill, Douglas McArthur, Fiorello LaGuardia, and Teddy Roosevelt. On the spiritual side, I was inspired by Sir Thomas More (in particular the Thomas Bolt play and later movie by the same title, *A Man for All Seasons*). A close second was Francis of Assisi, as depicted in the Zefirelli movie, *Brother Sun, Sister Moon.*

A female leader who inspired me was Joan of Arc; I was particularly enthralled by the extraordinary narrative of her contained in Churchill's four-volume treatise, *History of the English-Speaking Peoples.*

It would be impossible to understand the historical figure of Joan of Arc without recourse to some element of spiritual inspiration. The same is true for Gandhi, Lincoln, and Dr. Martin Luther King.

They were clearly inspired by some inner force within their own psyches; and they evidently passed on the inspiration to others, often in a way that has little rational explanation. Their impact begs for an analysis such as the previously cited work by Swart, Chisholm and Brown, titled *Neuroscience for Leadership*).

The trio explain that "charismatic leaders both feel and elicit powerful attachment emotions for their cause, their people, their wider community."

A lot has been written about the kind of "charisma" that allow leaders to elicit passion for their "cause" and their "people." Not as much, except in political biographies and in their own letters and speeches, has been written about what moves the leaders to possess those "attachment emotions" in the first place.

Here I can best analyze myself; in simple terms, the parable that moves me is the one Jesus told about the "talents."

Faith-Based Motivation: the Parable of the Talents

One motivation drives me more than any other; it is the idea that there are others in my community who don't have the basic necessities of life. I recognize that I inherited physiological gifts like athletic and scholastic ability. And I am inspired by a biblical passage that suggests that will some day have to account for how we used those inherited "talents."

As per the prior quote from Victor Frankl, I have a strong sense of "what life expects from me." Let me expand briefly on that.

One day, when I was 11 years old, my home was taken over by communist soldiers; they had removed my dad and older sisters to temporary prisons, as the government anticipated what later turned out to be the Bay of Pigs Invasion. It was at that moment that I made up my mind to be a political leader.

Later on in life, I sensed that I had most of the traits ("talents," to use the biblical term) necessary for leadership – except for one.

One trait I did not inherit. They call it "charisma."

In the age of television, the single most important trait for a politician is charisma. And I compounded the deficiency by choosing engineering as my initial profession.

Engineers compete with accountants to win the prize for lack of charisma. For me, the modest bit of charisma I display came later, and at great exertion.

I had to learn charisma.

Charisma and the "Myth of Powerful Men"

The father of the science of sociology is thought to be the German sociologist Max Weber. Borrowing from Weber for a recent *LA Times* article, Virginia Heffernan explains that the charismatic leader is distinguished from:

> *...both rational leaders, who derive their legitimacy from their office and accomplishments, and traditional leaders, who are 'to the manor-born' monarchs and patriarchs.*

The reference to "manor-born" leaders has little applicability today, since we no longer have monarchs or patriarchs, except in the Middle East and in some European nations in which the king is a figurehead. The more pertinent distinction made by Weber is the one between a rational leader, who governs strictly based on the rule of law, and the charismatic leader, who isn't obeyed "by virtue of tradition or statute, but because they believe in him."

Even more pertinent is the combination of the two traits (rational and charismatic) in a leader. That was the case with American presidents John F. Kennedy and Ronald Reagan, as well as Canadian Prime Minister Pierre Elliott Trudeau.

Closer to home, a brewing exemplar of that combo is my own son, Francis X. Suarez, who is currently mayor of the City of Miami. I bring him up not to brag, but because I have intimate knowledge of him – both genetically and circumstantially.

Francis is the oldest of my four offspring. In some measure, all four combine their mother's and father's traits.

But charisma by itself is like "ganas" without inspiration, or talent without motivation. You need a strong dose of fearlessness to overcome the kinds of obstacles that have confronted great leaders.

Fearlessness is a trait that Winston Churchill shares with iconic leaders like the ones previously mentioned. Modern sociobiologists, who render little or no credit to a person's sheer willpower, would characterize seemingly fearless leaders as classic "alpha males." But that obviously doesn't apply to Gandhi, let alone Joan of Arc. Lincoln never struck me as a classic alpha male; nor did Dr. Martin Luther King.

Sir Thomas More was anything but, if you believe the various biographies. He was meek in the face of threats, sensitive when confronted by the weakness of others, and forgiving towards the treachery of those he had previously trusted.

In summary, the kind of fearlessness we see in the previously mentioned leaders is something that can only be associated with faith in a greater power. It is a trait that is unique to our species and can be defined as the ability to set aside our own welfare and subsume it to the welfare of our entire species.

It is fortuitous that a handful of leaders, as well as a multitude of lesser-known, but equally or more heroic souls, have listened to our better angels at key moments in history. The faith-based persistence of such great leaders, coupled with the sheer goodwill of a multitude of good, ordinary people, has saved our species time and again.

They rightfully lay claim to being on the right side of history.

And now it's time to discuss other, more ordinary (less well known) personalities. One of those – with which I have a great deal of intimate experience – is my wife. She is charismatic, in the sense of being the quintessential "people person."

My wife doesn't have a leading role in society as a whole; she is simply the natural-born leader in any gathering of family and friends.

Perhaps being of Cuban descent helps; it is not an exaggeration to say that it is part of our national character to be musical, boisterous, and talkative.

The "Loudest Voice"

My wife Rita is the classic case of someone who is a charismatic non-leader. When I say "non-leader," I mean that she doesn't want or seek a leadership role in society as a whole.

In smaller groups, she is very much a leader – in the sense of being the life of the party. The best vignette of that trait is what happens at the beach in the summer. The role she plays is more evident on the rare days when she doesn't show.

Let me set the stage.

We have a group of friends who spend their summer at the beach.

There are bankers, doctors, teachers, homemakers and every other kind of occupation. For the most part, they are in their fifties, sixties and seventies. But there are younger ones too. Typically, the younger ones are the second generation of our peers.

My wife Rita is a unique member of this tribe. She is the entertainer, the spokesperson, the "loudest voice." The best way to describe the role she has in the beach group is to describe what happens when I walk alone to the beach. On those rare days when she is absent and I am present, it's as if the oxygen has been removed from the group.

"Omigod," they exclaim, "where is Rita?" When I announce that she's not coming, it's a serious shock to the system. You can see the sadness creep into their eyes. Rita is, indeed, the life of the party. She makes the rest of the group laugh, and cry and cringe at the near-death incidents in her life.

She is the "loudest voice," the uniter, the Pied Piper, the leader of the band.

She is the spice of life. The question is: Why do humans need other humans to provide spice to their lives? Perhaps it helps to analyze the opposite side of the spectrum personality: the one that most people belong to.

At the extreme opposite of the charisma spectrum from my wife and my son is the bland personality.

The Bland Personality

On the opposite side of my wife's ebullient character is the run-of-the-mill member of our species. Among our beach friends, of which there are about 25, as many as 15 are run-of-the mill. They don't need to make themselves heard or noticed.

They are perfect examples of why our species thrives on the company of others. To be happy, to be fulfilled, we need to socialize – to communicate and *entertain or be entertained*. It can be in the form of conversation, or art, or movies or plays. It can involve music. It can include prayer and some form of religious ritual.

Based on my experience, most people would prefer to be entertained than to be the entertainer. Most people are happy not to be the protagonist in the play. They toil in private, mind their own business, and stay in their shell.

That works most of the time in most human endeavors, where the willingness to follow the lead of the boss, or obey norms set by the leader, is of tantamount importance. One area of human activity where that "blandness" doesn't work is sports.

A lot of anthropologists believe that sports is a modern invention needed to replace the fight-to-the-death that humans experienced when they transitioned from the evolving species of ancestors, who were (in their opinion) advanced results of natural selection.

Whether that's true or not, what is definitely true is that athletic competition evinces a lot of elements of survival of the fittest. But there is a difference. Most sports that people like and follow and pay to watch are team sports. (Tennis and golf are exceptions.)

My own favorite sport is basketball. At the college level, it is currently played best by my *alma mater*, Villanova University.

In basketball, as in other team sports, the best team is not always the one that has the best individual performers. It is not simply the sum of its parts. Some teams play better than one would expect by simple, quantitative analysis.

Some say it is a matter of "attitude" more than pure, athletic skill. And that brings me to a book with that title.

"Attitude" at Villanova

In modern day America, athletic competition is the stuff of life – more than the fluff of life. It was said, during Roman times, that the emperor would keep his subjects quiet and happy by giving them "bread and circuses." There is some truth to that in modern, American life.

At present in America, the highest level of athletic competition is the professional level – the NBA in basketball, for example. The second highest level is the collegiate one, involving universities like Villanova.

Perhaps no program is as successful, in basketball at least, as Villanova's. I reckon the reason is a great tradition for excellence in that sport, coupled with a cloak of spirituality which covers and infuses the varsity basketball team, the players, the coach and the religious chaplain.

In a book (titled *Attitude*), Villanova's coach, Jay Wright, explains the mantra that envelopes Villanova's success. What we should draw from it is an important lesson in human psychology.

Here's the magic of Villanova's success. Most of the players in a program like Villanova's are raised in the inner cities of America. They are typically from working-class families; a majority are African-American.

A substantial percentage of Division I basketball players see the possibility of a pro basketball career as their best chance to move up the economic ladder. Schools like Duke, Kentucky, North Carolina, U.C.L.A. and Indiana regularly send a handful of their best players to the pros. Villanova does too, but one gets the impression that they are slightly less athletically endowed and substantially more well-rounded, more mature as human beings. Schools like Notre Dame and Stanford also have that reputation.

Unlike Notre Dame and Stanford, Villanova is small in enrollment and endowment. It has a certain regional reputation, but not a national following. And yet, it is exceptional in many ways.

Two of Villanova's athletic programs are world-class. One is its track program, which for many years was coached by the nation's most celebrated track coach, "Jumbo" Jim Elliott, who worked for free. (Which is another example of altruism....)

How Villanova has prevailed in a contest that involves well over 200 universities, many of which are state-run and state-funded, is a wonder of wonders. There are two key factors that distinguish Villanova basketball: one is selflessness and the other is a certain resiliency that results from being inspired by the team chaplain. The chaplain, Fr. Bob Hagan imparts on all the players a long-term, spiritual view of life's challenges.

In addition to drawing inspiration from the team chaplain, Villanova players are surrounded by an artificial, extended family of graduates who have played and excelled. Perhaps 10% or 15% made it to the professional ranks. Whether they are famous or rich, or athletic journeymen who made it in business or teaching or law or medicine, the alumni extend a network of support to the current players. If a player is struggling, or gets hurt (ruining his chances to turn professional), the entire network of coaches, chaplains, and alumni envelops that player with their advice and their resources.

150

In other words, *Villanova basketball is a family affair.* And the inspiration, the spiritual fuel, is provided by the Augustinian ethos.

St. Augustine is known in the secular academy as the writer of *The City of God*, a treatise which highlights the necessary connection between church and state. (Among religious academicians, he is known for the introspective, intrinsically remorseful, and – to me, at least – enormously boring *Confessions*.)

Perhaps better than any philosopher in history, and certainly earlier than any of the well known ones, St. Augustine represents the fusing together of the two fountains of truth: reason and revelation. As previously mentioned he is perhaps best known for his dictum: "Believe, so that you may understand; understand so that you may believe."

In other words, science and religion must work together to craft the totality of human wellness.

Villanova stands for that. And its basketball team fuses together the best insights of science with the best insights of faith. Individual aspirations are subsumed into the success of the team – not because they represent a lesser objective, but because they lead to collective success.

The right "attitude" is like a blanket, a covering, a common denominator that unites each player to his teammate. It's yet another example of synergy – of the phenomenon by which the whole is greater than the sum of its parts.

Tribal Synergy: Roseto, PA

The micro example of Villanova basketball plays out in two much larger sociological models. One is a little town in Pennsylvania and the other is the enormous demographic group to which we refer as Hispanics or Latinos.

The little town in Pennsylvania is named Rosetto and is located in Central Pennsylvania; its story comes to us by virtue of Malcom Gladwell, who wrote about it in his famous book, *The Outliers*. Gladwell narrates that a couple of doctors – one local and one not – were talking one day about how healthy the people of Rosetto were. The local doctor mentioned to his friend, who was visiting from Oklahoma, that people in Rosetto seldom committed suicide, or suffered from clinical depression or addiction.

The two doctors engaged a sociologist (John Bruhn) to do a statistical study in an effort to determine what the reason for such local wellness was. Here's how Bruhn described the peculiar townspeople:

> *There was no suicide, no alcoholism, no drug addiction and very little crime. They didn't have anyone on welfare. Then we looked at peptic ulcers. They didn't have any of those either. These people were dying of old age. That's it.*

Gladwell explained that when Bruhn presented his findings to the medical community, they were met with great skepticism. But he stuck by his findings, including the correlation of collective psycho-social health to a sense of belonging and of cultural-religious commonality of values.

Roseto's demograpics were quite similar to two other Pennsylvania cities, i.e., Bangor and Nazareth. Both cities were about the same size as Roseto and were populated by what Gladwell describes as "the same kind of hardworking European immigrants."

But those two cities did not have the same commonality of culture and religion as Roseto. Gladwell describes a town in which the people all "went to mass at Our Lady of Mount Carmel and saw the unifying and calming effect of the church." In a town of barely 2,000 people, there were twenty-two civic associations. And the people shared a "particular egalitarian ethos," which "discouraged the wealthy from flaunting their success and helped the unsuccessful obscure their failures."

The coherence of a small town cannot often be replicated in the wider society. But it's instructive to see how a combination of faith-based egalitarianism can help those prone to various mental ailments avoid or lessen their morbidity.

Another, broader population of Americans has been recently studied and has shown a similar egalitarian ethos; it is also at least partially faith-based.

I refer to Hispanic-Americans.

Learning From Hispanic Americans

In recent article by syndicated journalist Nicholas Kristof, the author refers to a "Hispanic Paradox." The paradox lies in the fact that "despite poverty and discrimination, Hispanic Americans live significantly longer than white or black Americans." And Kristof expands on the unusually positive sociological indices that Hispanic-Americans represent:

> *Latinos also appear to have lower suicide rates than whites, are less likely to drink alcohol, are less likely to die from drug overdoses and, at least among immigrants, appear to commit fewer crimes....It is a paradox because the disadvantaged normally live shorter lives. Hispanics in the United States endure discrimination, high poverty, lower rates of health insurance than both whites and blacks – yet they enjoy a life expectancy of 81.8 years, compared with 78.5 years for whites and 74.9 years for blacks.*

By way of explanation, Kristof refers to various scholarly studies, including a recent one that tackled the condition of Hispanics in various cities that showed how Hispanic-Americans displayed a special "resilience" when dealing with the coronavirus crisis. And he concludes:

> *Part of the explanation may be that what many white Americans think of as "traditional American values" – an emphasis on faith, family and community ties – are disproportionately found among Latino immigrants...*

Note the trio of factors: "faith, family and community ties." In effect, what the studies indicate is quite supportive of the theme of this book. Faith, family, and community ties, based on an egalitarian ethos, are the variables that make for a civilized and equitable society.

Interestingly, science has been reaching the same conclusion in dealing with a broad range of mental-health conditions. Some call it holistic medicine; it could just as easily be called "synergistic cures."

Synergistic Cures/Holistic Healing

By the end of the twentieth century and the beginning of the twenty-first, the therapeutic health sciences found themselves looking not just for explanations, but for cures, outside of the strictly biological.

They went from calling it "health" to calling it "wellness." The emphasis shifted from talking about chemically correct medicines to talking about "holistic medicine" or "cure for the soul."

Mental health is perhaps the area of medicine in which the person requires more than a biological examination and consequent cure. And within mental health, the different forms of addiction range from the chemical effect of a drug (which in the case of alcohol reportedly affects about 4% of all Americans) to a totally different phenomenon, initially not chemically addictive but which becomes addictive when the repetitive choice of a substance or sexual experience is wilfully abused.

In the next chapter I explore, in particular, the addiction that starts off as narcissism and ends up as sociopathic. I conclude that such an addiction, because it has no initial biological component, is quite close to what theologians refer to as "sin." It requires the kind of self-control that can only result when the subject acknowledges the dictates of his own conscience – of "right reason."

It is akin to what Catholics do in confession.

By contrast, most addictions have a biological or psychological component. They result from voids or defects in the nature or nurture of a child. Often, they lie latent and are brought to the surface by a drug or substance, which can be recreational or medical.

The classic addiction, and perhaps the most common, is to alcohol or drugs. The cure often includes group therapy that goes by the name of the "Twelve Steps," of which the last one is discussed in the next chapter.

As with other cures of the soul, the cure of drug and alcohol addiction often requires a strong dose of social and spiritual inducement. In this case, it is not so much that the subject has to acknowledge his or her sinfulness, as much as a connection to others who are in the same straits and also (much to the surprise of agnostic health scientists) to an infinite source of unconditional love that can supply the self-esteem that is lacking.

CHAPTER X
ADDICTION AND GRACE

With our backgrounds in psychiatry, psychology and education, we authors have all witnessed that motivation and addiction are on a spectrum. This seems obvious when it comes to drugs and alcohol: you are motivated to seek something that makes you feel good, but at some point it is no longer good for you but you still keep doing it. Tara Swartz, Kitty Chisholm and Paul Brown, *The Neuroscience of Leadership.*

$\mathcal{S}$ome human failings, which we tend to characterize as being in the realm of "mental health" are best understood only in light of what the above authors call a "spectrum" and is really more of a matrix.

Let me explain. Human emotions interact with the human power of choice to dictate our conduct; when the will is in control, it limits the use of alcohol or medicinal drugs. A person takes a pain pill for the necessary period of time only. Once the pain is lessened or eliminated, there is no more need for pain medicine.

The same is true for alcohol, or sex. In the proper measures, alcohol and sex can enhance the human experience, motivating the mind-psyche to greater pleasure, bonding and creativity.

However, at the edge of the control spectrum, where the person is incapable of discipline, often for biochemical reasons, the power of decision-making is lost. The person becomes a slave to the intoxicating stimulus of the drug or sexual activity.

What modern psychology is beginning to realize is that the human spirit can be an important factor reinforcing the will, which in turn leads to powering the will. The spiritual inspiration (referred to in theology as "grace") does not substitute the mind-psyche, nor does it "cure" the biochemical reaction of the pleasure-giving drug; but modern science increasingly lends credence to the idea that it is a real, contributing factor in the equation.

Often, it is transformative.

On the Addictive Personality and Its Treatment

The treatment of substance or sexual addiction has long been considered one that requires strong components of mind and spirit, in order to stimulate the necessary measure of mental control.

For that reason, I have previously referred to that condition as "numino-psycho-somatic." This three-dimensional approach is gaining acceptance in the popular and scientific literature.

For years, the spiritual component has been considered effective in the treatment of substance abuse and addiction. It is in that field that I came across the writings of Dr. Patrick May, whose book gave me the title for this chapter: "Addiction and Grace."

Put simply, the idea is this. When you try to cure an addiction, be it to alcohol, drugs or sex, it is typically necessary for the therapy not only to extract the negative component, but to replace it with a positive component.

Dr. May illustrates by using a physiological analogy. He explains that often, when a tumor is removed, the surgeon fills the void left from the removed tumor with some inert liquid that occupies entirely the place previously inhabited by cancerous cells.

This technique helps to reduce the chances that the cancerous tumor will regrow. Interestingly, it has an equivalent in Japanese literature.

The idea that removal of negative elements should be replaced with positive ones is conveyed by a bit of Japanese poetry, which goes by the name of "Kintsukuroi" or "Kintsugi." Literally, the word means "to repair with gold." Here's how it's described in the literature:

> When the Japanese mend broken objects, they aggrandize the damage by filling the crack with gold. They believe that when something's suffered damage and has a history it becomes more beautiful.

There is also a well-accepted theological equivalent of that poetic metaphor, but first I want to discuss the more empirical side of addiction.

How Widespread Is Substance Addiction

In the United States, it is estimated that at least four percent of all adults suffer from substance addiction. Until recent times, the great majority of those were those whom we call "alcoholics."

Perhaps no American leader – and certainly no American president – was more popularly known as a "drunkard" than Ulysess S. Grant. Reading Ron Chernow's superb biography of Grant in the light of modern science, one can easily grasp the reality that alcoholism is at least partly biological.

As Chernow writes, "alcoholism is a 'chronic disease,' not a 'personal failing' as it has been viewed by many." For Grant, "alcohol is not a recreation selfishly indulged, but a forbidden impulse against which he struggled most of his life."

That distinction is key, and very present for me, as I suspect I am in the former category – of those who use/abuse alcohol as a "recreation selfishly indulged." In other words, I can stop drinking any time I apply even a little bit of self-control.

Not so for the four percent of Americans whose genetic composition predisposes them to fall into daily, constant, impossible-to-stop abuse of alcohol. I know some intimately. Unlike me, they cannot have a drink or two and then have dinner and forget all about the alcohol, enjoying fully the sense of a full stomach, followed perhaps by a good night's sleep.

For those poor folks, substance addiction is truly a chronic disease. And the hope is that it can be cured by giving up alcohol altogether and attending sobriety meetings of the kind we associate with the 12-step process of Alcoholics Anonymous.

Without delving into the success rates of such group-therapy programs, it can safely be said that for a distinct minority of our species, their alcoholic tendencies have, at least initially, *little to do with bad choices and much to do with bad genes.*

But even as to those, a strong dose of spirituality can help enormously to overcome the illness. Although they fell in the trap based on physiological characteristics, the way out is partly guided, and enormously stimulated by the thought that there is an all-loving being who prepares for them a better life, in which such pleasures can be enjoyed in moderation – and without fear of addiction.

Modern psychology is learning a lot from the ancient wisdom that gives spirituality an important role in human wellness. One senses a departure from the very secular beginnings of the science, under its first and best-known exponent.

From Freudian Psychology to Numinosity

From its very beginnings, psychology was seen as a totally secular science in which the idea of God was, at best, a nice, self-acquired trait – a crutch to carry our species through times of material or emotional deprivation.

More recently, psychologists – and particularly socio-psychologists who study group behavior – are discovering what is for most of us an obvious nexus: that which exists between faith in a creator and psychic health. There is even an archive of scholarly studies and articles dedicated to the topic.

It is called the "Archive for the Psychology of Religion." And it contains a multiplicity of well researched papers. One such "Research Report" was just published in 2019; it was entitled "Evaluating the relationships among religion, social virtues, and meaning in life." Three co-authors are listed: Neal Krause from the University of Michigan, Peter C. Hill from Biola University and Gail Ironson from the University of Miami.

Having examined an extensive bibliography, the authors readily conclude that "a number of different facets of religious life have been linked with meaning making." That conclusion is followed by an impressive list of examples:

> *For example, some investigators propose that a sense of meaning arises from participation in religious rituals, including worship services. In contrast, other researchers maintain that meaning is found in the process of developing a deep sense of commitment to religion, while yet other scholars argue that meaning emerges by adopting security-focused religious beliefs (e.g., the belief that God watches over the faithful). In addition to this research, some investigators report that positive emotions are associated with a deeper sense of meaning in religious settings, while other researchers demonstrate that that people who are more likely to take the perspective of others and simulate their thoughts, feelings and intentions (i.e."mentalize") are more likely to derive a greater sense of meaning in religious contexts.* [Citations omitted.]

The authors then proceed to identify "beliefs that are shared by every major faith tradition in the world." They categorize them as three key "social virtues" which are roughly defined as follows: compassion, forgiveness of others and a willingness to provide "social support for others."

This listing is almost exactly the one that I have previously identified as the three social virtues that are most clearly faith-based, except that I combine compassion with the willingness to provide support for others and call it "faith-based altruism." Forgiveness of others is what I call "faith-based tolerance."

Those are clearly the most important traits taught to us by both ancient religions and modern science.

But let's discuss, for a moment, a third trait, which is actually more of an ego-centric one. I have called it "faith-based persistence" and suggested that it is the one particularly needed for the alpha member of the species.

Persistence is key to any human engagement, whether it be sports or work, which requires overcoming failure, fatigue, and fear. In other words, pretty much every human activity that is intensively productive or competitive.

Clearly, that trait is essential to a the leader. (In ancient societies, it was essential, for survival, to the entire species....)

Interestingly, the drive to succeed, the *ganas* of math teacher Jaime Escalante (featured in the movie "Stand and Deliver"); the *attitude* of the successful basketball program (featured in the book by Villanova's basketball coach, Jay Wright, by that name); the trait that distinguishes Gladwell's *Outlier*, requires a strong dose of self-confidence.

One could easily confuse it with narcissism.

Excessive amounts of this trait can be harmful not only to the individual, but to those around her. Its treatment often requires a serious dose of the spiritual.

As we shall see, too much self-love is psychologically damaging. Too little self-love is psychologically debilitating.

To some extent, the cure for both is the infinite love of an almighty being. The God ingredient in psychoanalysis has gained acceptance, particularly among practitioners of the healing sciences, as opposed to the theoreticians.

A very strong correlation has been found between the idea of a God and the yearning that most in our species have for a more perfect form of happiness, both on this earth and in a possible afterlife.

Numinosity = the Power of the Spirit

I have a psychiatrist friend, Dr. Francisco Maderal, who is Cuban-born and U.S. educated. He is quite erudite, having gone through the rigorous training in languages, philosophy and theology that all Jesuits must complete before actually becoming priests. Having entered the seminary at 18, he was not formally ordained until he turned 32.

Later in life, Dr. Maderal left the priesthood and got married. In between, he completed his studies in psychiatric medicine. He has a busy practice in Miami.

Once I posed the following question to Dr. Maderal: "How important is the spiritual component in the cure of psychological ailments?" He answered with a term I had never heard. "It's all combined," he said; "I refer to the holistic treatment of the patient as being numino-psycho-somatic."

I was familiar with psychosomatic phenomena, of course; and I also happened to remember, from my Latin studies in high school, that the root "numina" refers to the soul/spirit. But I had never heard the three terms combined into one word.

Here is how psychiatrist Thomas Moore defines the spiritual component (numina):

> *Religion specialists have a word for the power of a spiritual presence – 'numinosity.' The word comes from the Latin 'numen, having a strong spiritual quality or suggesting the presence of a divinity.'*

The idea that there is a divine being, and that such a being cares for the ailing patient is quite powerful – the more so as one gets away from the purely physical ailment (say a broken bone or a tooth that has decayed) and moves towards the realm of mental illnesses.

I was fascinated when I read a book that explained the spiritual dimension in connection with the treatment of bipolar depression. This condition, formerly known as manic depression, is thought to be incurable even by modern medications, which can lessen the cycles of depression but cannot eliminate them.

Let's delve a little bit into this revealing book.

"New Hope for People with Bipolar Disorder"

Although I am not an expert on bipolar disorder, I have some experience in dealing with this affliction on the part of a handful of family members and close friends. I have also read extensively about it.

One of the most interesting and insightful books is the one with the above title, written by a trio that includes the patient, the psychiatrist and the rabbi (Jan Fawcett, M.D., Bernard Golden, Ph.D. and Nancy Rosenfeld).

Describing what modern medical science says about bipolar disorder, the authors emphasize that there is consensus that the illness is almost entirely, if not entirely, biological:

> *Bipolar illness has long been a source of fascination in the mental health field. The fascination stems from the*

161

dramatically different states of mania and depression, which
are alternating yet integral components of the same illness....
[B]ipolar disorder is arguably the most biological of the
mental illnesses...[and yet] the psychological environment
appears to activate the genetic vulnerability converting it into
a life-long illness.

The patient herself stresses that even though the biological phenomenon (the cyclical bouts of mania and depression) is not totally curable, a psycho-spiritual component can help greatly alleviate its effects. She quotes a passage from rabbi Harold Kushner (author of *Why Bad Things Happen to Good People)* in which he explains his own illness in a theological context: "God," he says, "gave me the strength and wisdom to take my personal sorrow and forge it into an instrument of redemption which would help others."

Applying the theological wisdom to herself, she says:

Even acceptance of our own mortality enhances the meaning of
life and, as Kushner reflected, gives each of us the opportunity
to be productive and to impact others so we'll be remembered
as having contributed to life.

At the opposite end of the volitional spectrum is the condition referred to as "sociopathic" behavior. The sociopath appears to have a much greater component of volition – of pure choice to feed an inclination we all have towards selfishness. This is a trait that nowadays goes by the scientific name of "narcissism."

The person who only responds to the instinct to survive is called by psychologists a narcissist. There is a fine line between the narcissist, the sinner, and the sociopath.

The Narcissist, the Sinner and the Sociopath

Any modern psychologist will say, without hesitation, that all humans are narcissists, to some extent or another. Similarly, any Protestant minister, Jewish rabbi or Catholic priest will say that all humans are sinners.

Looking introspectively at myself, both pronouncements ring true. We are all narcissists; we are all egotists; we are all sinners.

Clearly, it's a question of degree. It's not too much different from the distinction we draw, in legal philosophy, between the occasional transgressor and the hardened criminal or "recidivist."

Catholics divide the seriousness of sin into two classes: venial or mortal. Venial is not catastrophic, but can become so if not corrected; mortal is debilitating and must be erased with contrition and the sacrament of confession. Those are theological concepts; and in this book we try to limit ourselves to purely scientific concepts, avoiding metaphysics as much as possible.

But the parallels are worth analyzing. Moreover, it seems that psychology and theology merge when one analyzes the sociopathic personality. Let me elaborate.

In the sociopath, the psyche and the will seem to merge in a pathological way, producing a personality disorder that seems to have equal parts volitional and psychological.

People suffering from that personality disorder, which evolves from excessive, uncontrolled narcissism, are categorized by psychological science as sociopaths. In effect, the recidivist narcissist becomes a sociopath. In that sense it is very similar to what Christians call the "unrepentant sinner" and what psychologists call the "pathological liar."

Scientists who have studied carefully the behavior of sociopaths tell us that it is a condition in which the person ceases to know the difference between falsehood and truth. The narcissistic impulse is so strong and so repeated that it distorts reality.

In its extreme manifestation, the narcissit-turned-sociopath ceases to have compassion for the victim of her narcissistic excesses. Because the reality is distorted, the sociopath feels no remorse. Even when the lie is discovered, the sociopath justifies it by every twisted rationalization imaginable.

One therapist described a classic case. It involved a mother who justified taking a child out of a school where he was thriving and putting him in a school where he failed his courses. To the therapist, the mother justified her actions with lies about the child's performance, saying he was struggling at the prior school, when in reality he was doing quite well.

Later, it became evident that the mother did not want to drive her son to the first school because it was much farther.

I saw this phenomenon in my own family. Because I had thirteen brothers and sisters, I was able to grasp a correlation between no less than three incidents in which my mother fell into the same kind of sociopathic behavior.

I am not saying that my mother was a sociopath – among other things because I have no credentials to diagnose her behavior. Nor am I judging her in the moral sense; psychologists aver that the extreme narcissist/ sociopath acts that way due to a defense mechanism.

Here's how the previously cited author, Russ Federman, explains it:

> *The implications of narcissism have little to do with negative social traits. The narcissist is just displaying a broad range of adaptations that at one time felt necessary for his or her psychological survival.*

Psychologists tend not to be judgmental. It goes against the grain of that science to place blame. I suspect it is because psychology, like anthropology, is skeptical of the idea that humans can ever be fully free to choose to do good or evil.

Philosophers and theologians take a different approach. For the most part, they buy into the notion that humans can be virtuous or wicked. Virtue consists in never choosing to do harm to others in order to benefit oneself. Wickedness is simply the opposite. The wicked will step over their fellow beings to pursue their own wellbeing.

That metaphor describes the behavior of a narcissistic mother whom I never met, but whose conduct was narrated to me by her son, who happens to be not only my godfather but my intellectual mentor.

My Godfather's Mother

This narrative is based purely on hearsay. I have no way of confirming the veracity of it. However, I have no reason to doubt it.

My godfather was named Ignacio Warner. He was brilliant, enormously well read, and as objective in his judgments as any person I ever met. We called him "Tio Warner."

He had no reason to be overly critical of his mother.

His story went as follows. It happened that Tio Warner suffered a stroke and was lying on the ground, awaiting transport to the hospital by the fire-rescue unit.

His mom was apprehensive, but not for the reasons one would expect. She was not well dressed; had no make-up; felt a little less than attractive.

So what did she do? In her eagerness to get to the bathroom and change her outfit and her appearance, she literally walked over the prone body of her son. At that particular moment, her narcissistic impulses kicked in. She cared not one whit for her son; she cared for her appearance.

Again, I don't want to be too harsh in my judgment. Tio Warner's mom, for all I know, was simply taking advantage of the moment. She was not hurting her son in any way; she was simply ignoring his plight.

Perhaps no harm was done.

That is not the case with the prototypical, narcissist mom.

The Narcissist Mom

I previously referred to cases involving a mother's naricissism and how it often involves the choice of schools for her children. Psychologists confront the classic case, in which a mother is concerned about the child's performance in one school and suddenly transfers the child to another school, where the child's academic and social performance suffers.

It takes a while for the psychologist to dig out the truth of the matter, which turns out to be precisely the opposite of what the mother claims. On further investigation from the schools concerned (or confrontation of the mother with the psychologist's suspicion), the truth emerges: The mother made the decision to transfer the child to a school closer to her, so as to reduce drive time.

In other cases, the tuition is less at the new school.

The chronic deception by the mother in these cases is a defense mechanism used by the mom to hide her narcissistic decision. If repeated enough, it converts her into a sociopath, who feels no remorse when lying; in the meantime, the child suffers from low esteem.

The choice of schools, which happens later in life for children, when they turn six or seven, is not as traumatic as the mother's projection of trauma to her kids when losing a child in the womb.

My Sister Margaret: Insufficient Nurture

My sister Margaret was the fifth of our fourteen siblings. She bore more than her share of suffering during her childhood. She passed away a couple of years ago, having lived most of her adult life separated from the rest of the siblings by geography and religion.

Before she died, she confided in me how much she suffered from thinking that she had somehow been responsible for our mom's miscarriage of what would have been the 15th sibling.

The story we pieced together – and is mostly confirmed by her older sisters – began when, as teen-agers, they attended a mass celebrated with a tone of defiance towards the Fidel Castro regime. I should explain that in the first two years of Castro's regime, the Catholic Church offered some mild resistance by convening the faithful and allowing them – at the end of mass – to chant things like "Cuba Si; Rusia No."

In this particular demonstration, when the mass ended, the crowd remained outside and continued their sloganeering. Castro's militiamen responded by firing their weapons in the air. When told of this, and that my sister was in the crowd chanting, my mom claims to have "felt her fetus kick for the last time."

This story stayed with my sister for most of her life as a sort of stigma, in that she should have avoided such a dangerous gathering, which caused my mother to lose her baby. To make matters worse, the narrative continued, by both accounts, to say that when my sister got home, my mother was too distraught to come out of her room.

None of this makes any sense. Women don't lose their baby because another, grown daughter may have had a dangerous scrape. And, even if they did, the correlation in this particular case is totally backwards: What we know is that the baby kicked in the womb, which meant the fetus was alive, not dead or dying.

Moreover, the more telling fact here is that my mom did not come out of her room to embrace and nurture the daughter. Furthermore, no mother blames a daughter for her miscarriage, even if it were true.

This incident was not the only one that plagued Margaret's early years. My sister endured poverty, exile, a frustrated religious vocation, and a minority condition in terms of gender, sexual preference, and physical attributes.

On top of that, she opted for military service during one of the most frustrating, horrific, and psychologically traumatic confrontations in American history – the Vietnam War.

The great Lutheran theologian, Dietrich Bonhoeffer, once said that he no longer "judged people by what they said, or even what they did, but by how much they suffered."

Judged by that standard, my sister Margaret was an admirable creature.

Moreover, her life, like those of many other totally altruistic people, puts the lie to any notion that our species exists with only one goal: to survive long enough to procreate. To the contrary, my sister is yet another, poignant and proximate example of the many men and women who choose freely to devote their lives to serving others to the point of giving up sex and even risking her life.

Margaret was much shorter than the rest of us; but the little person that she was had a very big heart. There is no way, using the methodology of natural selection, to even begin to analyze the worth of my sister Margaret. In the animal world, Margaret would not be a star.

The same is true of my wife's cousin. The fact that this lady is so naturally kind and generous puts the lie to the concept of original sin, or original predisposition to flout the natural order of things, as Jews and Christian believe.

There are many people who meet that description. It seems like they were born kind, without sin, without selfishness. It seems like they have original innocence. That brings me to Francis of Assisi.

Original Innocence

"In our obsession of original sin, we too often forget original innocence." So said Pope Innocent III to Francis of Assisi, as portrayed in Zefirelli's "Brother Sun, Sister Moon."

I can understand a person being so mature, so inspired by a higher, nobler being and so sure that altruism in this life begets an eternity of infinite joy, that she wills herself to selflessness. What's hard to understand is the number of wives, mothers, aunties and grandmothers who seem to live for others effortlessly, as if they had no narcissistic impulses whatsoever.

I call them "natural saints." I have written about this phenomenon in other books, including most recently in *Science and the Theory of God.*

One of the people I describe is a plain, ordinary housewife and mom. She is no Joan of Arc or Indira Gandhi. She is one of those people you cannot find in a Google search.

Not even once.

Her name is Maria Elena Haramboure (we call her "Manena"), and she has raised three children, a handful of grandchildren and at least one great-grand child who is now fatherless. To each of her children, grandchildren, great-grandchildren and spouses or partners, she has provided food, a home and sometimes her own bed.

Manena has also taken in stray animals and refugees from Castro's Cuba who had nowhere else to go. She has provided a refuge to both family members and strangers. In exchange, she has asked for nothing but a reciprocity of kindness.

And even that she often does not get.

Why does she do it? Does she have hidden vices? Does she have secret pleasures? Is she just more saintly than the rest of us?

As far as I can tell, Manena is just a naturally kind person. She was born in a country (Cuba), in a province (Las Villas), in a town (Sagua la Grande) in which humanity was at peace with nature and with itself.

Small towns in Cuba, before Castro-communism, were havens for the working middle class. They were very similar to small town, U.S.A. even now. Everyone had a job; everyone had a roof over their head; everyone had a family to support and protect them. Everyone obeyed the laws. Everyone married someone they loved, had as many children as God gave them and taught them the Ten Commandments.

Whatever vanity Manena had was wiped out by the exile condition. As a refugee in Miami, she has never had much; but she does have a house and what is called a "reverse mortgage," which means she has drawn out the equity in her duplex and now will pass no value to the next generation.

But she has taken care of them. And when I say that, I mean some major dysfunctions in her children and their spouses. She has been the rock of Gibraltar to three generations.

Manena seems to have been spared the stain of original sin. Or maybe she has it in smaller quantities than the rest of us. She is physically talented: a beautiful face and a beautiful voice. Her children and grandchildren are also quite attractive.

If the world were populated by 100% people like Manena, we would not need police officers, or jails or therapists.

Alas, most people have more vanity, or more lack of self-esteem, or more anger, or more lust for power or sex.

Maybe I should not quantify it. Maybe it's not most people. Maybe there are more Manenas and less narcissists than we realize.

It's a good thing, because the number of people born with special needs, plus the number who suffer developmental dysfunctions, plus the number who fall into some addiction or another is quite large. Luckily, the maternal and paternal instinct, coupled with the inspiration of a benign, almighty being, offers enormous support to the especially needy child.

I have a grandchild that was born with a strange, chromosomal deletion: one of millions of molecules that are simply out of place, or missing entirely from the billion-molecule equation that signifies human perfection. She is not able to speak or think clearly about what is happening around her. She probably senses the love that surrounds her, but cannot contribute to the material wellbeing of others – or even herself.

There are five kids in the family and the other four are as perfect as I have ever seen: they are cute, intelligent, super-well behaved, and loving.

They have nature and nurture aplenty.

And just as importantly, they have faith.

Life is bound to bring challenges to the other four in the family of five. There will be heart-aches, failures, broken bones, broken relationships, maybe broken marriages.

But the family is very united, very loving, very supportive. The parents emphasize the good traits in each child, including the one with special needs. Chances are that the four "normal" children will have their own mates and their own offspring, and that there will be a small percentage of those with physical or mental deficiencies. When those come along, the important thing is to care for the soul as well as the body.

For most of us, the care of our body is a consuming endeavor. Most of us dread being hungry; we plan assiduously to have the food necessary for our sustenance, a cup of wine with our meals (or a cocktail before them), warm clothes and a soft bed.

We also – as a rule – want to live forever.

The last of these instincts, the yearning to live, has its exceptions in my immediate family. One sister who passed away recently had no compelling desire to live much longer and no desire at all to use "extraordinary means" (such as an open heart surgery) to survive.

One brother, who is close to me in age, recently revealed that he was ready for "Jesus's embrace" in what he expects will be a paradise involving basketball, good music and the love of family and friends.

I myself am not particularly ready for Jesus's embrace – at least for a few years, and then a few years after that. I follow the example of most senior citizens in America, who collectively spend about a trillion dollars each year to survive the last six months of life.

Each of the above insights is useful in understanding the totality of the human person. In philosophy, that kind of analysis is called "integral humanism." For a while, the Western world seemed to reject that approach, in favor or a totally materialistic one; I have referred to it as "scientific materialism," though it has also been called "scientism."

The intelligentsia in nineteenth century England (and to a lesser extent in the United States) seemed disposed to abandon such holistic analysis of our universe and rely only on empirical science.

Then came the "natural theologians."

On Natural Theology

Natural theology was made popular by various British thinkers of the nineteenth century. The most prominent was Bill Paley and his most memorable analogy was that of a watch that falls by the wayside, where it is found by a passerby.

Nature is very much like a watch. It has extraordinarily matching parts and if they weren't so well matched, the entire assembly would not work.

That well organized assembly of parts, if found by a passerby on the side of a road, would clearly reflect a watchmaker. It would not be thought by any reasonable person to be a random assembly of parts that happened by accident. The same is true of the proverbial hurricane whose winds pass by a junkyard and carry a fully assembled car in its wake.

Things just don't happen that way – not without an intelligent craftsman to create them. It is even more obvious when you consider living organisms.

The same way that the human body seems to be crafted by an extraordinarily creative intellect, the human psyche seems similarly crafted. Just as the body needs physical nourishment, the psyche needs psychic and spiritual nourishment.

Art, music, literature and film all nourish the psyche. So do human relationships, including and especially those powered by romantic love. All psychologists understand that.

But there is one human trait that needs to be viewed through the prism of the spirit. It is the ability to forgive.

I have already referred to the mental illness of clinical depression, and how its treatment can be enhanced, in many cases, by the virtue of forgiveness.

Forgiveness As a Psycho-Spiritual Trait

If I had to state which is the single most important contribution to psychology made by the historical figure of Jesus, it would be the importance of forgiveness. And not just of others, but of ourselves.

Catholics have a sacrament we call "confession" or "reconciliation." It is not too different from the "conversion" that Protestants speak of, although for Catholics it's a routine thing and not a once-in-a-lifetime phenomenon. It is also similar to the yearly "atonement" of Jews.

Speaking in strictly secular vein, Lee Iacocca (in his marvelous autobiography) talks about the psychological value of confession. He explains how psychologically useful it is to examine yourself and then follow that introspective exercise by articulating to another human being where you have failed to be charitable, or humble or diligent in carrying out your duties.

Seeking forgiveness from others, and from God, is a useful exercise. Forgiveness cleanses the soul and fills the psyche with joyful energy. It is particularly important in a marriage.

Married couples, if they are successful in their union, sense this bit of wisdom, perhaps by instinct or more likely by trial and error. We know that when things get hot, we need to be the first to apologize and caress and prostrate ourselves before the spouse, even if only 1% at fault.

And it's safe to try that at home.

It is also safe to try among neighbors, tribes and nations. The advent of Christianity can be seen as the beginning of reconciliation between warring nations. It would take a whole book to catalog the importance of forgiveness after a bitter war – such as the American Civil War and World War II.

Again, American exceptionalism comes to mind. In the case of WWII, the Marshall Plan, whose total import is estimated to have been about 50 billion in today's dollars, is arguably history's best (maybe the only) example of a powerful, victorious nation coming to the aid of the losing transgressor, as well as their foreign victims.

It says here that it could not have happened were it not for the power of faith. And I also suspect, although I cannot prove it with scientific tools, that the more we forgive others for their transgressions, the more we will be forgiven in the other life – assuming there is one.

And so we are ready for the last chapter, which *does assume that our species has some role to play in eternity.*

There is no scientific way to prove that. At this point we travel from the world of physics to the world of metaphysics – from the world of deductive and inductive reason to the world of pure deduction. There is no way to test what I offer by way of other-worldly insights, except by what Jacques Maritain calls "knowledge by connaturality," which is a fancy way of referring to those little voices inside us that reflect hidden realities inside our psyches.

It is the only part of this book that requires a leap of faith. But my reader should know that it is a *logical leap of* faith; there is no superstition involved. In that sense, it is no more speculative than the equations of relativity or thermodynamics, as they are extrapolated back to the Big Bang.

And I want to add – parenthetically – that my last chapter is more scientifically reliable than the entire field of Superstring Theory or the previously described "Panspermia" as an explanation of biogenesis. Moreover, my reflections on the after-life are more scientific, more tangible, more explainable in terms of the human psyche than the entire phenomenon of Dark Energy, which is nothing less than 70% of all the mass and energy in the universe.

And which is still unexplained. Perhaps unexplainable.

All of those mysteries of science, inscrutable as they are, constitute what scientists grapple with on a daily basis. So far, they have been stymied in the effort to put forth a scientific "Theory of Everything."

In that sense, this book is more advanced – in terms of connecting the dots of what we know about our species – than what the best theoretical physicists at Princeton can offer. And Princeton is where Einstein and John Forbes Nash taught.

CHAPTER XI
UNTIL THE END OF TIME

Man is the only being that knows death; all others become old, but with a consciousness wholly limited to the moment which must seem to them eternal, that instills the essential human fear in the presence of death." Oswald Spengler, *Decline of the West.*

Man is literally split into two: he has an awareness of his own splendid uniqueness in that he sticks out of nature with a towering majesty, and yet he goes back into the ground a few feet in order blindly and dumbly to rot and disappear forever. Ernest Becker, Cultural Anthropologist.

$\mathcal{J}$ found these two quotes in a brand new book by the renowned Columbia University professor of physics and mathematics, Brian Greene. The new book is titled "Until the End of Time," and follows on best sellers like *The Elegant Universe* and *The Fabric of the Cosmos.*

Like me, Greene bases his anthropological analysis on math and physics, as well as other so-called "hard sciences." We who are engineers and who study physics, are bound to notice that there is a special human curiosity when it comes to the beginning as well as the end of time.

We might as well admit that we are obsessed with how our lives will end and whether there is an after-life.

Others are content to live life as it were a permanent condition. Then

174

they are hit with the tragic death of a loved one, or an unexpected illness that brings them to the brink of death.

Some of us are so content with our lot that we don't dwell on death at all, until our twilight years – and even then, not too much as long as we *have a quantitatively ambiguous yet qualitatively valued few years of life left.*

That all changes when a sure death is imminent.

The tragedy of the so-called 9/11 terrorist attack in America quickly changed perceptions – particularly for the few, inside an airplane controlled by terrorists, who looked at death in the eye and accepted it.

In a recent anniversary of the 9/11 terrorist attack, and using her inimitable style, Peggy Noonan gave us a glimpse of what our species is capable of thinking, saying and doing when we know our earthly life is ending. This circumstance, as much as any that can be envisioned, brings out the spiritual dimension in humans.

In the precious minutes between the realization that they were about to die and the crash itself, various passengers in United Flight 93 (which crashed in Pennsylvania) sent brief messages to their loved ones. A typical one was by flight attendant Ceecee Lyles; she left a message to her husband on his answering machine. It said: "Please tell my children that I love them very much. I'm sorry baby, I wish that I could see your face again."

Note the themes. First and foremost, the woman faces death with total resignation. No anger or bitterness. Secondly, her thoughts are not for the suffering she is about to endure, but for the suffering that her children will endure. Last but not least, she conveys unconditional, unrequited love.

Others left similar messages. Brian Sweeney "called his wife, got the answering machine and told her they'd been hijacked." His message was: "Hopefully, I'll talk to you again, but if not, have a good life. I know I'll see you again."

The most famous call was from Tom Burnett, who was organizing a defensive maneuver that he assumed would get him killed by the terrorists, even if others were saved from the carnage intended by them. He said: "We're all going to die, but three of us are going to do something." And to his wife: "I love you honey."

Noonan summarizes what all these people were saying:

These people were saying, essentially, 'In spite of my imminent death, my thoughts are on you and on love. I asked a psychiatrist the other day for his thoughts, and he said the people on the planes and on the towers were 'accepting the inevitable' and taking care of 'unfinished business. At death's door people pass on a responsibility – 'Tell Billy I never stopped loving him and forgave him long ago.' 'Take care of mom.' 'Pray for me, Father, Pray for me, I haven't been very good.' They address what needs doing.

Peggy Noonan says that in spite of imminent death, people "address what needs doing." But that is really short-changing our fellow human beings. What these people were addressing was what would give meaning to their shortened lives, by giving worth to the much longer lives of others.

In other words, what needs doing for love to prevail among our peers. Humans find meaning in helping others, even if they are total strangers. It has nothing to do with survival of the species – let alone their own survival, which is now precluded.

This is not convincing to many of the anthropologists. To them, the whole concept of an eternal after-life is a concoction of our own mind-psyches, as needed to cope with life's vicissitudes. In his 2015 book, *The Instruction of Imagination,* for example, Daniel Dor argues that the terror of knowing we are going to die "rendered our ancestors quivering piles of biological protoplasm on the fast track to oblivion." To deal with that terror, suggests the previously cited Ernest Becker, what saved our ancestors "was the promise of life beyond physical death, either literal or symbolic."

In other words, our species concocted the whole concept of an afterlife, substituted the fear of death with the yearning for life with an all-loving, all-knowing God, and invented "faith."

That conclusion doesn't even begin to ring true to me. What rings true to me is belief in an infinite being who crafted a universe with remarkable precision, allowed one creature the power to reject Him/Her, observed the chaos the ensued that rejection, and decided to intervene once again in human affairs to restore harmony. Viewed from the perspective of history and both the empirical sciences and the social sciences, the Judeo-Christian paradigm fits the evidence rather well.

To me, and hopefully to my reader, the conclusions of integral humanism are more a leap of logic than a leap of faith. (Or maybe, as previously stated, a "logical leap of faith.")

Hopefully, my reader will have reached the same conclusion by now, based on the data and the arguments put forth.

And now I end my message by telling the story of a wedding, involving my nephew (who is like a son to me), his bride and a magical evening in Miami, where everything seemed right with the world.

The Wedding of My Nephew

A young man who is very close to me got married recently. The occasion brought into town half of my remaining siblings, of which there are nine.

It also brought together people from all walks of life, including the very rich, the classic undocumented immigrant, and a fairly diverse group of law enforcement officers.

Before I get to the officers, I want to single out two extremes in the socio-economic realm. In the United States, the extremes interact at social events like never in history, in any part of the world.

Parenthetically, Britons who visited America, during the colonial period, were shocked that Americans of all economic and social levels shook hands, as equals, when they met! That was rare 250 years ago, but is more or less the norm throughout the world today, with the possible exception of India and the Middle East; and in Europe for royal families.

The U.S. is the quintessential, egalitarian country. At the wedding we could see it with two extremes who were represented there. One, who is the employer of the groom, is a billionaire born in Iran. The other, who has been befriended by the groom's father, is a Romanian refugee who is struggling to obtain residency in America, as he makes a living with musical gigs.

A poor, eastern European refugee meets the billionaire from Persia. In between the extremes, every profession and occupation was represented, from a mayor and a congressman, to entrepreneurs, lawyers, firefighters, teachers (like my wife), retirees and the aforementioned police officers – to whom we now return.

The law enforcement officers were mostly male and mostly married to younger women, who were in most cases, quite attractive. They were classic cases of alpha-males, married to alpha-females.

Police officers have a very high rate of divorce. The reasons given include the unusual hours and the temptation, for the alpha male, that comes with authority.

For each case of a divorced policeman who marries a younger, sexier woman, there is a strong believer whose marriage lasts the duration. At social events, my wife of 40+ years and I often gravitate towards the solid family men and women when deciding with whom we share bread. Then, when the music starts, we gravitate towards the younger set, who are often single or recently divorced. In a strange and wonderful sense, my wife and I mix better with our kids' generation than our own.

That is not the case for many of my contemporaries. It's hard for me to identify with most of them. When compared to couples our age, my wife an I are odd ducks. All of my peers, it seems, are retired or desperate to retire.

I feel like I am only halfway through my self-education and perhaps two-thirds through my expected political life, which began at 36 and might end at 86, or even 96. Besides this book I am finishing two others, which will bring my total to 8. Eventually, I expect to write somewhere between 10 and 15.

I am not ready to die. But it is not because I loathe the idea of eternity, as some writers have portrayed it.

Edgar Allan Poe/Tennessee Williams on Death

For some of our greatest writers, the idea that our species has a limited lifetime is devoid of any romantic flavor whatsoever. When mourning a premature death, Tennessee Williams exclaimed:

> I shrieked with horror; I plunged my nails into my thighs and wounded them; the coffin was soaked in my blood; and by tearing the wooden sides of my prison with the same maniacal feeling I lacerated my fingers and wore the nails to the quick, soon becoming motionless with exhaustion.

That's about as macabre as one can put it. Herman Melville is not much less despondent about our fate, reckoning that even when rough waters seem to have subsided:

> All are born with halters round their necks; but it is only when caught in the swift, sudden turn of death, that mortals realize the silent, subtle, ever present-perils of life.

Part of the problem with some of these authors is that they don't have a very favorable view of the after-life. Here's Dostoyevsky (through one of his characters, bearing the exotic and unpronounceable name of Arkady Svidrigaylov) on the circumstances that might await us after death:

> Eternity is always presented to us as an idea that we can't grasp, as something enormous, enormous! Why does it have to be enormous? All of a sudden, instead of all that, imagine there'll be a little room, something like a country bathhouse, sooty, with spiders in all the corners, and that's the whole of eternity...

It's not convincing. It's not logical. It's not what our Judeo-Christian culture is about. God loves us too much for that kind of ending.

What Is God's Plan for Us?

You don't have to be a Christian, or a Jew or a Muslim to conclude that the Big Banger/Creator/God is an exceedingly loving being. The logic is simple.

You start by looking at nature, at the Big Bang itself, at the amazing music that our species composes, at the rainbow, or the *aurora borealis*, or a snow-capped peak, or any single film by the Discovery Channel.

Alternatively, you review the latest findings of astrophysics, quantum physics, the inflationary moment, and biogenesis. There is an infinitely intelligent and infinitely powerful being out there. All of this didn't happen by accident – among other things because it would violate the laws of probability. (Scientists use the fancier term – the Second Law of Thermodynamics....)

And by a huge factor.

So there is a God out there, whether Nietszche likes it or not. (I love the poster that someone concocted that read: "God is dead" – Nietszche. Underneath that it read: "Nietszche is dead" – God.)

Empirical science in the first half of the twentieth century, and therapeutic science in the second half, joined together to give us a holistic understanding of our species. That much is clear from an objective understanding of both history and science.

And we are on the threshold of proving, with some degree of scientific certainty, that there is a God, and that the more we learn and follow Her/Him, the better we will thrive, both individually and collectively. (A good book to read is by Robert J. Spitzer, titled *New Proofs for the Existence of God: Contributions of Contemporary Physics and Philosophy.*)

And that is where science, alone, leaves us. We now have a handful of different scientific arguments for proving the likely existence of God. Add those to the five given to us by Thomas Aquinas.

So the question becomes: If there is a God, and he/she is infinite in time and power, why create beings with the ability to reject God, violate nature, mess up the works?

Well, there is only one logical explanation: Because God loves us.

God certainly doesn't need us.

And why does God allow bad things to happen to us? Well, because he/she made us so similar to him/her that we have the power to either be loving or not-so-loving.

You can add to that, if you want, a special feature probably possessed by an infinite being, who is self-sufficient by definition, and also loving. Such a being is probably a complex interaction of "persons" or entities that are eternally in loving union.

That concept is known as the "Trinity." It is not necessary for our analysis, but it sure is a nice fit: If our species is made in the image and likeness of God, then being triune (three-dimensional) might just be a useful insight that we initially derive from ancient writings and now confirm by all the science summarized here.

In any case, an all-loving God would hardly create a species that yearns for eternity, avoids death at all costs, and then crashes into the wall of joyless death. My law school classmate, John D. Hagen, Jr., puts it this way:

Every adult of a reflective turn of mind has thought to himself in situations of great joy: 'I am happy now, but not so happy as I ought to be because I know that this joyful time will soon be over. How I wish that, somehow, it could be prolonged indefinitely!' This yearning after the infinite is natural to all of us."

And Hagen concludes:

Man, thus, has a conscious capacity to experience immortality and a soul suited by nature to outlive death…It would be remarkable if God should place in man a conscious potential for lasting happiness, and then make it impossible that he fulfill it.

Although we cannot prove it, we can safely and logically conclude that some sort of paradise awaits us after death. That's the first important conclusion to draw.

The second conclusion follows logically from the first: it is also consistent with a proper reading of the history of our species, as we know the 5,000 years since Abraham. It is that humans who are caring and generous also tend to do good things.

As a result, things get better with time. Perhaps much better, even in this earthly life.

Made in United States
Orlando, FL
05 December 2022

25606591R00125